T0258944

Situational Project Management

The *Dynamics* of Success and Failure

Best Practices and Advances in Program Management Series

Series Editor
Ginger Levin

Situational Project Management

The *Dynamics* of Success and Failure

Oliver F. Lehmann

CRC Press
Taylor & Francis Group
Boca Raton   London   New York

CRC Press is an imprint of the
Taylor & Francis Group, an **informa** business
AN AUERBACH BOOK

CRC Press
Taylor & Francis Group
6000 Broken Sound Parkway NW, Suite 300
Boca Raton, FL 33487-2742

Printed on acid-free paper
Version Date: 20160524

International Standard Book Number-13: 978-1-4987-2261-2 (Hardback)

Visit the Taylor & Francis Web site at
http://www.taylorandfrancis.com

and the CRC Press Web site at
http://www.crcpress.com

Dedication

I dedicate this book to my wife Silvia, whose support for its creation was inestimable.

Situational Project Management
...as "navigating between monsters"

The Monsters:

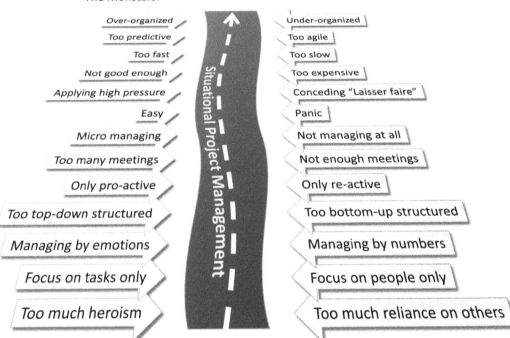

Contents

Foreword

Situational Project Management (SitPM) is a new concept recently introduced into the project management body of knowledge. Little has been written about SitPM, but what has been is fragmented, and a clear definition is yet to be put forth. Some have aligned SitPM to the Situational Leadership Model of Ken Blanchard, probably because of similar terminology; while others, such as myself, believe that it goes much deeper into process than it does into behavioral relationships. Its name is suggestive of project management approaches that are somehow adjusted to or accommodated to align with the nuances and changes in project situations. I have been advocating and writing about such adaptive approaches for nearly 20 years and offer a holistic definition of SitPM from the perspective of my own client experiences. I have felt like one crying in the wilderness until Oliver's book came to my attention.

Definition: Situational Project Management

Most project managers would agree that each project is unique. But not all project managers would agree that the best way to manage a unique project would also be unique. Many still cling to the old practice of having a methodology that is applied to all projects. "One size fits all" is still in common use, and this has proven to be a ticket to complex project failure.

Project uniqueness (i.e., the project situation) is defined by four factors:

- Physical characteristics that define the PROJECT
- Behavioral characteristics that define the project TEAM
- ENTERPRISE environment of the project
- MARKET environment receiving the deliverables

These factors establish the requirements that an effective project management must include to be successful. Because of the dynamic nature of all four of these factors, any chosen project management approach will not remain static across the life span of the project. It may be further adapted or replaced altogether.

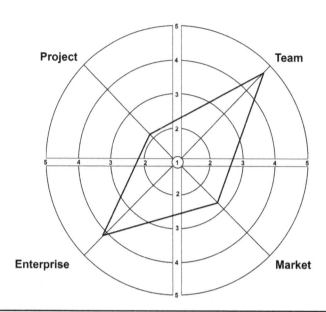

Figure a   The SitPM Spider Chart for a Specific Project

This definition provides a broadly based statement of what constitutes the SitPM Framework and a foundation on which to further develop the processes and practices of SitPM. There is much to be done, but we have a good start on that journey with this book.

The SitPM Framework can be envisioned using a graphic such as the one suggested in Figure a.

On a factor scale, 5 represents a highly complex and uncertain level of that factor. So, for this example, Enterprise and Team factor scores are high and require special attention as the specific SitPM Framework "recipe" for this project is crafted. Project, Enterprise, and Market are discussed in the ECPM Framework book cited at the end of this Foreword. Adding Team and its interaction with Project, Enterprise, and Market completes the SitPM Framework.

It would be a mistake to think that the SitPM Framework presents a Do-It-Yourself management approach. That would be courting chaos. Rather, to be effective, a SitPM recipe must be constrained within a portfolio of tools, templates, and processes. These must be vetted and designed for use specific to the culture and practices of each enterprise. Guidelines may be issued for their appropriate use. Deviations from those guidelines can often require validation by the project manager. Such a project management culture offers a controlled flexibility on the part of the project team and preserves the creativity needed by the project team to succeed in complex project management.

The situation with SitPM is complex and ever changing and does not follow predefined recipes. To be successful, project managers must be *thinking* project managers. This is not a casual statement. It points out a distinct change in what an effective project manager must be prepared to deliver. I use a cook/chef metaphor. There was a time when project managers were expected to follow a defined recipe—they must be a cook and use recipes developed by others and approved by their management. Those days are gone. Today's complex and uncertain project landscape requires a project manager to be creative and craft the best fit approach to the management of a project—they must be a chef and able to create their own recipes on a project-by-project basis. Flexibility and creativity are essential to deliver project success.

The portfolio mentioned above defines the "pantry" from which the project team creates the "recipe" it will use to manage its project. But as the definition states, that recipe can change due to the dynamic nature of the project and its internal and external environments. The SitPM Framework looks more like art than science, and that is as it should be in order to succeed in the complex project landscape.

To effectively integrate the SitPM Framework into the complex project landscape presents several pre-conditions for the enterprise. I have my own thoughts on this, as will anyone else you might ask. Here are mine, and then I will share how Oliver's version aligns with them.

- **Establishing a Supportive Infrastructure.** This must be primarily supportive but also include compliance and review functions.
- **Defining the Project Manager Position Family.** Project managers must manage projects incorporating meaningful client involvement.
- **Designing the Training of Project Managers.** The focus should be decision making and problem solving rather than rigid processes.
- **Creating the Certification of Project Managers.** Competency-based certification is required at more senior project manager levels.
- **Designing the Change Management Process.** A lean change process is central to efficient and effective solution discovery.
- **Implementing Strategic Portfolio Management.** The complex landscape needs maximum flexibility for effective resource management.

All six of these challenges bring new processes and practices to the complex project management environment. Many challenges will involve journeys into the unknown and require their own Agile project in order to converge on effective solutions that deliver expected business value.

1. **Establishing a Supportive Infrastructure.** The Project Management Office (PMO) has been the mainstay, but it tends to be more standards and compliance based. It struggles to be a meaningful partner. The PMO will not be effective as a support infrastructure for a SitPM Framework. Instead, a Project Support Office (PSO) should be established. Although it still has a role similar to the PMO, its major focus is support and flexibility, not standards and compliance.

2. **Defining the Project Manager Position Family.** The project manager's role and responsibility extends to both the process that creates the product or service and the product or service delivered. To that extent, the project manager is a multidisciplinary position that should include the five disciplines shown in Figure b.

3. **Designing the Training of Project Managers.** Having the skills necessary to detect, analyze, and solve challenging problems is essential. The vetted portfolio should be rich with tools, templates, and processes to assist the project team with these activities. This is a long-term effort, as is the training to support it. The training support should be mostly short-term courses of one day or less. Both instructor-led and online versions should be available.

4. **Creating the Certification of Project Managers.** One who is certified as a SitPM project manager must possess not only the process knowledge but also the competency to perform the processes. Knowledge-based certification programs are appropriate only at

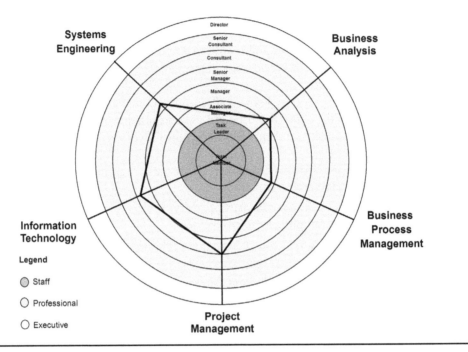

Figure b SitPM Project Manager Position Family

the lower skill levels. The IPMA Levels D, C, B, and A are the certification programs closest to what the SitPM Framework will need. It has both knowledge-based and competency-based components.

5. **Designing the Change Management Process.** An open change process must be an integral part of the SitPM Framework because change is the lifeblood of every complex project's search for an acceptable business solution. The major issue is to make solid business decisions regarding the integration of changes into solution development. The changes must not be instituted casually but rather as the result of due diligence and the overall view of the solution direction. The ECPM Framework bundled change management process is shown in Figure c.

6. **Implementing Strategic Portfolio Management.** The SitPM Framework has a definite role to play in the execution of the strategic plan of the enterprise. We think of that role as an enabler role, but it will go beyond that into a strategy formulation role as well. This is new and yet to be developed.

Putting It All Together

The SitPM Framework is as much process oriented as it is behavior oriented. My work, as captured in my book, *Effective Complex Project Management,* is more process oriented. SitPM is more behavior oriented. But the two are complementary and not the least bit contentious. Behaviors will drive the choice of PMLC models and how they might be adapted to embrace specific behaviors.

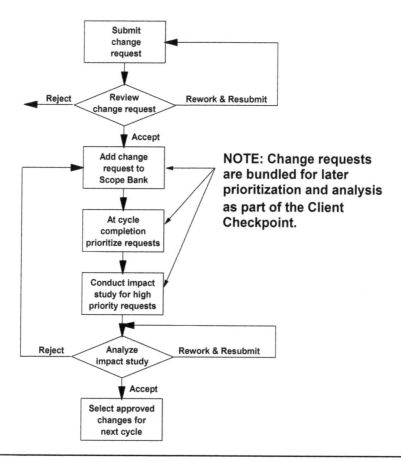

Figure c   ECPM Framework Bundled Change Management Process

Oliver's book is the next step in our journey of learning and discovery of the SitPM Framework. But it is a necessary step, nonetheless. He acknowledges many of the six challenges and offers a change-based perspective because that is what drives these complex projects. We have much to learn and discover about codifying the framework for SitPM, and Oliver has joined the journey. He has my hearty congratulations for having made a significant contribution to the project management body of knowledge. We will be good travelling companions!

Robert K. Wysocki, PhD
President, EII Publications, LLC

Robert Wysocki is the author of 25 books on project management, including the following, most recently published:

Effective Complex Project Management: An Adaptive Agile Framework for Delivering Business Value. Plantation, FL, USA: J. Ross Publishing, 2014.

Effective Project Management: Traditional, Agile, Extreme, 7th Ed., New York, NY, USA: John Wiley & Sons, 2014.

Both are unique in their contribution to the SitPM Framework.

Preface

Sometimes, things just fall into place.

In project management, we see many practices that have been developed, each of which has specific strengths and weaknesses. For some of them, we do not even know who invented them and when, such as the bar chart or the Work Breakdown Structure (WBS). Others have been published with the names of their inventors, such as the Critical Path Method (CPM) or Scrum, and some of them even have some kind of copyright protection. Some practices have been developed as work-arounds to others, and a project manager in a later generation may no longer perceive them as makeshifts, but rather as the way things need to be done.

Some practices are just technical. Others are highly political. Discussions among promoters of these practices are often conducted with an incredible zeal to prove that their own method is the best, under all circumstances, thereby placing political positions over project management behaviors, methods, or tools. Project management practices are often linked with a person's business. As an expert on method XYZ, one can make a living as a trainer, consultant, book author, or, of course, as a project manager. The fanaticism that these promoters of practices communicate may be a fire burning inside of them. Sometimes, it is an attempt to increase the business value of their own knowledge and skills that the customer or employer is prepared to pay for and to remain competitive against other experts with their distinct methods.

Project management practices are often inexpensive excuses:

- "No, we will not accept the change request. It may bring benefits, but it is not in the plan or the strategy."
- "No, we do not have a plan and we will not waste time documenting the project. We are going Agile."

Many practitioners in project management observe these discussions with irritation. Practices are often promoted in a way that taking sides with one of them requires the fundamental rejection of another one. Project managers are sometimes unsettled by this and wonder what they are going to lose with such a step. And they are right. The moment may come when this other behavior, method, or tool is more favorable in a specific project situation. It is helpful, therefore, to have a basic concept—what it is that makes a practice favorable or detrimental to a given situation and how this practice should be implemented. This is the concept that allows things to fall into place. This concept has a name: "Situational Project Management" or, in short, SitPM.

SitPM is not another practice for project management. It is rather an exploration into various project situations and practices. Each of the practices has been developed to respond to specific requirements and ignores others. Many were created while someone actually carried out a project; others were developed by experts with a certain type of project in mind. Most of these new practices are then promoted to be applicable to "all kinds of projects", and no limitations of its applicability are communicated and discussed. While the requirements placed on the project manager and the team are frequently changing, so are the practices that match the situation. It is then advisable to know various practices and have a guideline available to use in a specific project or project situation.

Looking at the many methods, tools, and behaviors that we can apply, there are definitively more project management practices than what a single person can master. As a project manager, one must be selective in the practices that one learns and applies. Some practices are required in a given industry or organization more often than others. Over time, however, one can add more practices to his or her portfolio of skills and will then be able to respond to challenges in a more situational fashion. It will then be interesting to see things fall into place. One may also be able to build teams with colleagues and friends who have mastered practices other than his or her own, and the combination of such competences and strengths can be a powerful approach to project management.

In this book, I hope to provide project managers with as much guidance as possible to develop this adaptive approach to their projects. To make the best use of this book, I recommend doing some analyses from time to time on the project:

1. To what type does the project belong?

2. What degree of dependence and independence will the project manager have?

3. What is the planning horizon that project (a) requires and (b) allows for in the two dimensions of time and granularity?

4. To what leadership behaviors ("Achieving Styles") will the project respond most positively?

5. How much direction does the team need and value? How much self-organization can it conduct and will it appreciate?

6. Based on which assumptions have project decisions been made so far, and what happens if they prove wrong?

7. How much nonproductive time is left for the project manager to do organizational tasks?

8. How will the project implement its change request management process not only to allow it to engineer benefits and add value but also to protect the project from catastrophe?

9. What software should the project manager use?

This book addresses these very basic questions and offers advice on adjustments to common situations. I recommend not considering this as a definitive description of Situational Project Management. It is rather the "Open Sesame!" from *1001 Arabian Nights*, which will lead you into the cave of the thieves, where a treasure awaits the open-minded explorer, but also where new challenges and threats may occur.

Note: The author produced many pieces of art in this book using Corel Draw 6.0 and some of its clip-art components, with licenses to the author from Corel.

To all knowledge only a brief triumph is allotted between the two long periods in which it is condemned as paradoxical or disparaged as trivial. But life is short, and truth works far and lives long: let us speak the truth.

—Arthur Schopenhauer (abridged)

Acknowledgments

I am most grateful to my wife, Silvia; my children—Antje, Daniel, Sandrine, and Tizian; and my grandchildren, Amelie and Helena, for their patience with me when I was busy with this book and could not give them the time and attention that they deserved from me.

My special thanks also go to Dr. Ginger Levin, who believed in me right from the start and was a highly critical reviewer during the time of writing, as well as to John Wyzalek (my Editor at CRC Press), who shared his support and trust in me. I would also like to thank the production staff at DerryField Publishing Services—Theron Shreve, Marje Pollack, Mary Lemons, and Susan Culligan.

In addition, I am thankful to the many experts whom I've had the honor to meet in person, and whose thoughts and beliefs helped shape mine, including James R. Snyder, one of the founders of PMI; William R. Duncan, primary author of the original *PMBOK® Guide 1996* and the trainer for my PMP preparations; Patrick Weaver, who is one of the few experts I've met so far who shares my interest in the history of project management; and Cornelius Fichtner, who, during a day trip to San Diego with me, took the time to discuss the most basic concepts of this book and who inspired me to realize it. Thanks also go to Deanna Landers, Robert Monkhouse, and Kris Troukens who, together with other idealistic friends and colleagues, founded Project Managers Without Borders and opened the profession to a new degree of altruism, benevolence, and magnanimity.

I should also not forget to express my gratitude to Professor Jean Lipman-Blumen, who taught me the secrets of Connective Leadership. In the community of her followers, she is considered the rock star in leadership research, and I am one of these followers. She has a central place in this book and in my heart.

Several book authors were also influential in helping me develop the situational understanding of project management and of the dynamics of success and failure, including Aaron J. Shenhar, Dov Dvir, Eryn Meyer, Harold Kerzner, Jerry B. Harvey, Nassim Nicholas Taleb, Oliver Hart, Philip Rosenzweig, Frederic Vester, Hannah Arendt, Peter F. Drucker, and many more.

Two sections of this book are based on research for which I received immense support from Bernd Müller-Dautenheimer, Boris Reichenbächer, Gaby König, Herbert Gonder, Jan Christian Westheide, Jürgen Sturany, Martin Keil, Michael Krekeler, Michael Dickmann,

Patrick Schmid, Peter Corbat, Ralf Friedrich, Regina Wolf-Berleb, Sigrid Schubert, Stephan Koß, Uwe Vigenschow, and Werner Waldner. Without the help from these project management experts, I would not have been able to develop the knowledge for the development of SitPM—Situational Project Management—and of this book.

I am especially thankful to Robert K. Wysocki. The first English book that I read on project management was his *Effective Project Management,* first published in 1995, and it is a great honor to have him write the Foreword to my book.

I would further wish to express my gratitude to the Board of the PMI Munich Chamber and the over 70 other volunteers, who often had to do their work without the support of the President, and who have done a great and professional job. The same is true for the over 42,000 members of my learning group, "I want to be a PMP®", in the social network LinkedIn, where experts were prepared to respond to questions of learners when I was not available.

My last "Thank you" goes to the customers and students I have worked with over a period of more than two decades in the professional training business in industry and to a lesser degree in academia. They believed in me, but also challenged my understanding with often tough questions. They helped me question and validate the models and practices that were the contents of my seminars and evaluate them against their individual realities of day-by-day project management. They helped me understand better what the dynamics of success and failure in essence are: an ever new mixture of situational intelligence, luck, and merit.

Munich, 6 November 2015
Oliver F. Lehmann, M.Sc., PMP®, CLI-CA
Project Management Trainer

About the Author

Oliver F. Lehmann was born in 1957 in Stuttgart, Germany. He has studied Linguistics, Literature, and History at the University of Stuttgart and Project Management at the University of Liverpool, UK, where he holds a Master of Science Degree. He had practiced project management for more than 12 years, mostly for the automotive industry and related trades, when he decided to make a change and become a trainer, speaker, and author in 1995. Among his customers are international companies such as Airbus, DB Schenker Logistics, Microsoft, Olympus, and Deutsche Telekom, and he has had assignments as a trainer in Asia, Europe, and the United States. He works for several training providers and is also a Visiting Lecturer at the Technical University of Munich. Oliver Lehmann is frequently invited to speak at congresses and other events, where he motivates his audience to open up to the incredible diversity found in project management around the world, and to the new experiences that await discovery by project managers.

Among Oliver's diverse certifications is the PMP® certification from the Project Management Institute (PMI) in Newtown Square, PA (USA), which he obtained in the year 2001. In 2009, he acquired the prestigious CLI-CA (Certified Associate) certification issued by CLI, the Connective Leadership Institute in Pasadena, CA (USA), which is held by only a small number of people.

Oliver has been a member and volunteer at PMI since 1998 and currently serves as President of the PMI Munich Chapter. Between 2004 and 2006, he contributed as an analyst for troubled projects to PMI's *PM Network* magazine, for which he provided the monthly editorial on page 1 called "Launch", analyzing troubled projects all around the world.

Oliver F. Lehmann believes in three driving forces for personal improvement in project management: formal learning, experience, and observation. He considers the last one often overlooked, but important, because attentive project managers do not need to make all the errors themselves—they also learn from the errors of their colleagues.

Chapter 1

The Situational View on Project Management

1.1 Introductory Questions

The following questions are written in certification test style. They will provide you with an understanding of the contents of the following sections and the questions that are discussed within the text. Answer these questions before you read the section and then again once you have finished it to discover what you have learned. This Introductory Questions section appears at the beginning of each chapter of this book.

---

1. A construction company is building a neighborhood that consists of a hundred mostly identical houses. Each house is sold to a family and the leeway given for customizing it to personal wishes is limited to only a few traits, such as colors and selection of some standardized assembly parts (doors, windows, etc.).

   In this case, which statement is most accurate?

   a) The single house is perceived as a project by the construction company.
   b) The single house is perceived as a project by each family.
   c) The entire neighborhood is not perceived as a project by the construction company.
   d) The manufacturing activities of the assembly parts are perceived as projects.

---

2. The company you are working for identified an opportunity to win new customers by developing a new product and launching it on the market place. The opportunity has a short-term market window and it will not be easy to finish the development and marketing project in time.

    In this early stage, it is not quite clear what the new product will look like in future. What should be the first step?

    a) A project charter for the project should be developed that includes some objectives and rough specifications. This charter will be detailed and refined during the course and development of the project.

    b) The company should not hire a project manager and start the project before a detailed feasibility and impact assessment has been made and detailed product specifications have been defined.

    c) The company should not undergo the pressure situation and leave the market to competition. Projects should only be started for the development of goods that can be clearly identified right at the start.

    d) The company should assign the project to one of their functional departments and fully rely on their communication skills instead of building a dedicated project team.

3. When making an investment in a project, investors require compensation for which of the following?

    a) An amount equal to the invested capital plus a premium for the risk of loss.

    b) A return of investment of at least 3% over the national central bank prime rate.

    c) The risk-free rate of return plus a risk premium plus a premium for inflation.

    d) Sacrifice of immediate use of cash for consumption or other investments, plus possibly coverage of inflation and risk.

4. Projects are different from operations in which of the following aspects?

    a) Projects are limited by resources that may not be available in sufficient quantities at all times.

    b) Projects are performed to meet objectives or satisfy needs, or to create another kind of value.

    c) Projects should be performed following the cycle of Plan-Do-Check-Act.

    d) Projects are temporary endeavors and are unique by nature and definition.

5. Unidentified risks can later in the project _____.

    a) be alleviated by using advanced methods of qualitative and quantitative risk analysis.

    b) diminish the value of risk management measures and jeopardize the entire project.

    c) cause issues that can be sufficiently controlled by adding more resources to the project.

    d) be dealt with by setting clear objectives for all project stakeholders.

6. You have taken over a project, which brings together team members from different industries. What should you be acutely aware of?

   a) Certain special terms may be used in the different industries with diverse meanings, which can lead to misunderstandings.
   b) You need to apply pressure on team members from the other industries to learn and use your terminology and lingo.
   c) You should replace the team members from other industries with persons from your own technical domain.
   d) Different industries are generally no problem, as long as all team members come from the same regional culture.

1.2 The Purpose of This Book

This book is intended to support project managers to meet one of the most difficult tasks that they may face: Coping with changing requirements on the practices they implement and the results that they need to create with their teams. The word "practice" is used here for the application of specific approaches, tools, techniques, behaviors, and procedures with the intention to guide action and bring about desired results. Results can include any types of deliverables, such as products, services, knowledge, or any kind of enablement that the project is required to deliver. Most project managers have experienced during their profession that the same practice that has led to success in one situation can cause failure in another. Many articles on project management, as well as books, papers, lectures, and other sources, nevertheless promote so-called "best practices", a modern term for "cooking recipes" or "magic potions", assuming that one practice has been found and tested, and selling it as proven to fit all situations. Organizations also often commonly implement "best practices" as an element of their internal methodologies. Project managers also commonly tell their colleagues that they should use a specific practice XYZ because it was so successful in their project and they feel certain that it will be successful in other situations as well. Even organizations communicate from time to time that they have converted all their projects to a certain method, ignoring that this method may be a great approach for one project but inappropriate for another one.

A basic problem for project managers and for the organizations that are sponsoring these projects is the lack of understanding of how project situations can be different and what practices can be expected to be beneficial or detrimental. This understanding will be the focus of this book.

Most project managers are aware that they have to adapt their practices to their specific projects, and even during the course of the project, they may have to adjust these practices from time to time to changing situations and conditions. Projects are rollercoasters rather than smooth rides, and when project managers believe that they are perfectly sorted out for managing a great undertaking, some surprising events will remind them of the basic uncertainty that comes with any project. The understanding that practices in project management must fit situational demands is not new, but no help has been given to project managers so far to assess situations and base the selection of practices on the results of these assessments. Giving such help consists of two aspects: increasing the understanding of project managers on what types of projects and project situations commonly occur and giving hints for the favorable or detrimental effects that can be expected from the application of certain practices in these situations.

A project is like an exploratory journey, where each step forward creates new knowledge for the travelers, replacing uncertainty with certainty, risks with problems, and forecasts with actuals. While the journey starts with the need to make assumptions and base decisions on them, it ends with the opportunity to make assessments of what the decisions yielded. Figure 1-1 shows the development from uncertainty to certainty.

The book is partially based on a small Situational Project Management (SitPM) research project conducted during 2014 and 2015 with a group of 17 field experts. It is further based on the experience and observations collected during over 12 years practicing project management and over 20 years working as a trainer, most of the time with a focus on preparation seminars for the Project Management Professional (PMP®) certification in classroom environments. An interesting characteristic of these seminars, which deal with the diverse aspects of project management and the challenges to the organizations and people involved, is the cross-cutting view on the variability of project situations: the project managers who attend the classes come from different industries and business environments and have different personal experiences. A trainer may make a statement that is perfectly valid for one participant, but may not suit the professional setting of another one. When these differences become obvious, and when the trainer and students start digging deeper, the underlying differences become discernible and open for discussion. The statement "one size does not fit all" may sound banal and stereotyped, but it should nevertheless be a guiding principle for project managers and field experts.

How do experience and learned knowledge relate? Having actively practiced project management is important for any professional in the discipline—project management cannot simply be learned at a school and then applied. Practical experience helps one gain familiarity and awareness with the tools and techniques commonly used in project management and to understand

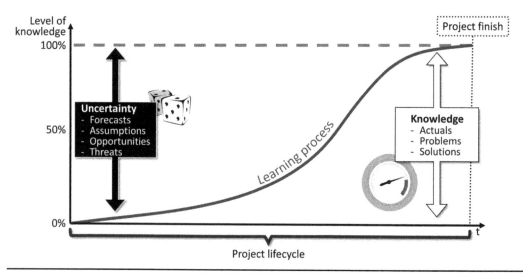

Figure 1-1   The beginning of a project is mostly characterized by uncertainty. Moving forward along the project lifecycle, uncertainty is replaced with knowledge.

their value. This personal contact with the discipline is important, but it can never be complete and will never suffice to understand and master all eventualities found in the complex environment in which project managers perform their jobs. For example, some projects are run as internal projects, which means that they are cost centers for the performing organizations that run them for their own purposes. Other projects are performed for paying customers under contract and constitute profit centers for the performing organization, the contractor. The experiences that project managers collect in either of these two types of projects are significantly different.

Most (not all) project managers in internal projects are in a weak position because their projects do not provide income. The projects' deliverables may do this later, but the projects themselves do not, and they are not performed against contractual requirements with paying customers. The requirements are instead put in place by internal requesters, which may be functional departments or major programs to which several projects contribute. Most project managers in customer projects are more powerful inside their own organization and have a better reputation. They champion the survival of the organization by producing a stream of income and must, at the same time, ensure that the organization does not run into a breach of contract situation with the project customer. Project managers often have no personal experience in both types of projects; their professional history is dealing solely with either customer projects or internal ones. Project managers running internal projects mostly consider themselves agents of change in organizational, technical, or other areas. Project managers in customer projects manage an existing business relationship (at least an important element of this relationship) and have to consider the often conflicting business interests of two parties: customer and contractor.

Project managers who wish to develop a full and professional understanding of project management have to add observation, empathy, and insight to their own experiences for project situations that they have not gone through thus far. Project managers may benefit from such a widened understanding beyond the limitations of their own experience. Different situations may turn up in the future, and it can be helpful for project managers to be prepared for them. Another reason to look over the fence between the two types of projects is the common need for cooperation: Customer projects on the contractor side often work for internal projects organized on the customer side, and problems on either side of the fence may impact the project on the other side.

The example of the two types of projects, internal and customer projects, will be discussed in more detail in Chapters 2 and 3, along with other types of projects. This book will recommend a project typology that may help project managers better understand and implement the situational needs on the practices that they apply.

1.3 A Primer on Project Management

Project management is an organizational discipline. Its importance in business, scientific research, arts, and politics has significantly increased over the last couple of decades. It is a discipline that focuses on one-time endeavors of a generally temporary and unique nature, in contrast to the operations discipline, which—in this understanding—pays attention to ongoing and repetitive activities. Along with the uniqueness of projects as complex endeavors comes uniqueness of the results of each project, which may be products, services, or other kinds of results. As stated previously, this uniqueness differs from operational activities, which are the ongoing daily business of most organizations that produce a recurring stream of identical or similar deliverables by numbers, often vast numbers. This stream allows the organization to develop repetitive routines, automation, and optimization. Operations can exploit scaling effects—assuming that the more items of a product or service that are produced, the less expensive the individual item or service will be for the manufacturer or service provider. Operations also have different learning curves, following the common experience that things that have been done repeatedly will be done in a more effective and efficient mode compared with things that are done for the first time. Another benefit of operations is that they are much less susceptible to personal issues of the people involved. For example, a lovesick streetcar operator will probably still be able to transport people safely along the tracks because most routine tasks do not require as much reflection, planning, and attention.

Projects, by definition, create results that are to some degree new and unique, and optimization and scaling are rarely an option. Personal issues can damage projects heavily, and conflicts among individuals or teams can make it impossible to be successful. Because of the rapidly changing environment of technology, economy, and society, organizations need to rely on effective project management as much as on efficient operations. Efficient operations help them develop predictability and order, whereas projects help them develop and implement new ideas, technologies, and forms of organizations, transforming their structures in a way necessary to meet future requirements. Project management also helps organizations comply with ever-changing regulatory requirements and exploit new opportunities for income.

While some projects are business-case driven and are performed to meet goals and objectives of the organizations, others are necessary to comply with law and other mandatory requirements. For example, from 2003 to 2005, corporations listed on US stock exchanges had to become compliant with the US Sarbanes-Oxley (SOX) Act. This law was enacted in 2002 after a series of fraud scandals in public corporations. It was designed to protect shareholders of corporations and required substantial investments by organizations listed on US stock exchanges. New systems had to be put in place to ensure that information given to shareholders was accurate, audited, and up to date, and they had to meet tight deadlines. Corporations listed on US stock exchanges had no choice but to run their SOX projects to meet strict legal requirements. Some companies used this occasion to create additional value from the projects that they needed to run. Some teams utilized this opportunity to renew aging systems and fix errors that had been resolved with wasteful workarounds. They replaced provisional arrangements—considered temporary when they were set up—with up-to-date solutions, increasing effectiveness and efficiency. These teams added business value to their projects, but the original intention to run these projects was not business value but a legal requirement—in essence, the sheer need of the organization to survive and the understandable wish of top managers to not have to go to jail, which was the final threat used by the US legislation to enforce the measures.

Mandatory projects are not rare at all: Companies all around the world have to meet laws for environmental friendliness, protection of personal data, financial transparency, protection of workers and consumers, and more, and the costs of these mandatory projects are enormous, while they may not yield any financial benefit to the mandating organization.

Voluntary or mandatory, projects are investments, and project managers are the administrators of these investments. The investments may be made to fulfill the interests of the organization and its investors, which is mostly true for internal projects and capital projects, or for the interests of other organizations under contract, mostly for paying customers.

Projects are temporary by nature. This may mean that a project has a clear starting point and also a clear end, but in real life, these moments are often blurred. Operations are also temporary; nothing lasts forever, but they are rarely intended to be temporary. Operations will try to run a business as long as possible, and the moment to finish it is rarely linked with the achievement of certain goals, but rather with wear and decay of resources, outdating of deliverables, changing requirements in the environment, insolvency of the performing organization, and similar factors. A central task of operations managers is to make their tasks and themselves indispensable. Project managers, in contrast, are expected to achieve specific goals and objectives, disband their team(s), finish the project, and thus make themselves redundant and free for the next project. Another major difference between project management and operations management is budgeting. Most operations managers have fiscal budgets that span the duration of a business year of the organization, and a short time before the end of a business year (or sometimes after it), the budget for the next business year is agreed upon between the manager and the organization. A small number of project managers also have to run their projects based on fiscal budgets. This can be a major problem because most projects must book resources early in advance, vendors must be taken under long-term contracts, and internal work packages must be assigned to business units of the organization. These assignments constitute longstanding commitments with financial impacts, and budgets are necessary in order to make these commitments. Projects rarely adhere to business or fiscal years; they may span two or more business years, and even a short project may be started in one fiscal year and finished in the next. A project manager who has to work based on a fiscal year budget would have difficulties entering commitments beyond the limitations of the current budget. The person would have to wait for the next fiscal year budget before resources could be booked and work ordered. While the person is still waiting for the budget to be approved, the resources may have been booked by someone else and the contractor or internal provider may have agreed to other assignments. Project managers often do not have a budget to manage at all; if they have one, it is a lifecycle budget that spans the intended duration of the project. Whereas project managers have no budget responsibility, the project budget will be managed by a functional manager, who mandates, supervises, and finances the project, a role often called the "Project Sponsor".

The interaction of projects with their organizational, social, economic, and environmental context is highly complex. While project managers are active agents of change and administer investments that increase the adaptability of organizations and their ability to innovate, at the same time, projects and the environments in which they are performed are also subject to change. Many requirements, objectives, and constraints that were valid at the onset of the project may not be valid when the project comes to an end, while others may have emerged during the course of the project, as business situations, intentions, and interests may have changed and as management strategies and approaches may have been adjusted. Other drivers of change during project lifecycles may be alterations of standards and laws as well as technical interfaces of items to which the project and its deliverables need to connect. Structures of organizations and people involved may also undergo changes while the project is run, often in an unpredictable way, sometimes benefiting the project, sometimes not. Another aspect of complexity is the ongoing interaction with so-called "stakeholders", persons, groups of people, and organizations who are potentially influenced by the project, intentionally or not, and may in turn influence the project. Stakeholders are those persons, groups, organizations, or entire states who need attention and who can demand implementation of certain changes on the project and its outcomes. While project management has long been considered an engineering discipline, skills such as identifying stakeholders and their needs, aligning the project to these needs, and engaging stakeholders (as far as this is possible) to actively support the project are commonly not considered technical. They are non-quantitative by nature and require interpersonal and organizational competencies. Looking at the interest that project managers take in stakeholder orientation, one can identify two types of project managers (in a wider sense):

Project engineers, who focus on the technical and functional aspects of a project, the "whats" and "hows" of the project; and project managers (in a more narrow sense) who also consider questions such as "With whom" and "For whom". The latter apply active stakeholder orientation in their projects.

Another facet of project management is that each project is a new learning process. The learning may include technical and economic aspects, as well as organizational and interpersonal ones. The learning process can include all stakeholders involved, including, of course, the project manager and a project management team, and good project managers stay prepared for lifelong learning. They build their professionalism on observations, knowledge, and experience that they have obtained in the past, but are prepared for the yet unchartered waters of future experiences, whose cruising still lies ahead of them.

1.4 Project Management Today

Project management is undergoing dramatic change, and while we are experiencing the process, no one can say for sure where it will lead the profession. Some examples of what can be observed are discussed in the following sections.

1.4.1 Speed of Change

The speed of change in and around projects is increasing. Old security-oriented, phase-gate approaches were great to stabilize processes and reduce risks. They are often mandated in organizational methodologies to control investment decisions and allow management influence on contracting decisions. But for many project situations, they have become too slow, and often, corporations cannot afford the time consumed by these approaches. Modern business environments require short time to market and rapid payback of investments. You may remember Hewlett Packard's (HP's) TouchPad tablet computer, a major product failure in August 2011. HP withdrew the device from sale only seven weeks after its market introduction. Although the market looked promising in 2010 when HP started development of the TouchPad, the market was no longer receptive when the TouchPad launched. HP's tablet product was receiving mostly positive tests by experts but could not cope with the matured "ecosystems" that had already grown in its competitors' gardens. Tablet computers need small software commodities that add value to them, called "Apps", and when the TouchPad launched, there were already over 100,000 apps available for Apple's iPad and 300,000 for Android devices[*]. Without a sufficient number of apps, the TouchPad was simply not in a position to compete: The shortage of apps reduced the product's value for customers, and without customers, software developers

---

[*] Numbers taken from statista.com.

were not motivated to create apps for the tablet computer. Unofficially communicated sales numbers were less than 10% of the original forecasts.

Projects that can be performed at a normal speed are becoming rare. Most people who sponsor them expect quick results. When you ask project managers in which aspect they feel most under pressure (meeting deadlines, delivering required functions and quality, avoiding unnecessary expenses, reducing operational disruptions, etc.), the typical response is time.

1.4.2 Open Skill Versus Closed Skill

The terms "open skill" and "closed skill" have been coined in sport psychology* but are also helpful in project management. Project management was long understood to be a closed-skill discipline—i.e., it is the project manager with his or her special skills in the discipline who makes the difference between success and failure. When you discuss the matter with practitioners and field experts, they mostly say that project management is an open-skill discipline. An example of a closed-skill discipline in sports is figure skating, where the success of the athletes is dependent on the person's training, discipline, body control, strength, and so on. The athletes must basically be able to mentally close their eyes to the environment and focus on their own performance during both preparation and presentation. There may be changes to the environment during the performance, but the figure skaters must ignore them and concentrate on their performances. They prepare by physically and mentally repeating all elements of the performance over and over again, until it is as near to perfection as possible, and the competition is in essence nothing more than another iteration of the same program.

Speed skaters, in contrast, perform a much more open-skill discipline. Most races are run by two concurrent skaters, and the athletes have to take into account the competitor, whom they want to outrun and with whom they are swapping the lanes once every lap. A speed skater racing too fast risks falling and getting hurt. If the athlete is too slow, he or she will lose the race. Speed skaters therefore observe their competitors and make decisions on their racing speed based not only on their own abilities, but also on the speed of the other skater.

Even more open-skilled is ice-hockey, where situations change in fractions of seconds, actions of team mates and competitors are often hard to predict, team compositions are frequently changed, and social interaction occurs, such as protection of the goal-tender by the defenders. A figure skater has a plan for each moment of the performance. Speed skating is more difficult to plan, because the skater is not alone on the track. For a hockey match, standard situations can be planned to some degree, but most of the time the players must be prepared to make decisions on the spur of the moment. See Figure 1-2 (on next page) on this continuum.

Most of the time, project managers do not perform a closed-skill discipline, such as figure skaters perfectly performing their programs, but need to act similarly to hockey players in ever-changing environments—often facing grim resistance. Brawling is much rarer than in ice-hockey, but it may even occur in projects. There may be moments in projects when the ability to perform closed-skill successfully may be an advantage, and when contemplation may help project managers deal with difficult situations, but most of the time, project management

---

* Some work has been done on the differences in how closed-skill and open-skill athletes prepare for their events (Highlen & Bennett, 1983).

Closed-skill

- Introspective
- Self-paced and self-controlled
- Competitive over the full tournament
- Focused on own performance
- Implementing a program
- Achieving perfection through seemingly endlesss iterations
- Managing known causes of risk

Open-skill

- Extraspective
- Self-paced and externally paced
- Competitive in any moment
- Focused on interaction
- Adapting to circumstances
- Achieving performance through situational intelligence
- Accepting known and unknown risk causes

Figure 1-2   Examples of closed-skill and open-skill disciplines: Skaters.

is an open-skill discipline. The world inside a project is ever changing, and so is the environment around the project. The same practice that was successful in one given moment may fail in another one. While introspection can give a project manager great insight about himself or herself from time to time, a project is won or lost through the interaction with its environment, its physical, functional, technical, or organizational aspects, but most of all through the care for its stakeholders.

1.4.3 Staged Deliveries and Multiple Deadlines

Most literature in project management presumes a single deliverable handover to an internal requester or an external customer. At the end of the project, a product has been finished, or the intended service-enablement or another kind of result has been achieved, and with the handover, the project gets finished and the usage of the deliverables begins. Some experts would allow for some short time after the handover to tidy up the development environment, to organize lessons learned and other types of documentation, and to finish additional nonproductive work before the next project can be started. This moment of delivery—handover, start of production (SoP), grand opening, or whatever its name—is then linked with a deadline: The results must be achieved at a certain point in time, and depending on the jurisdiction, forms of penalties or damage claims will be posed against the external contractor or the internal project team. Commonly ignored, another observation can often be made across all industries: Many projects hand over their deliverables in a sequence of stages. They do not have one deliverable transfer but many of them, so-called "staged deliveries", also referred to as "multiple handovers". Each delivery adds an increment to the previous one, thus further developing the product until it is completed in the last stage. Staged deliveries can be an emergency solution when a project is late, an attempt to pacify the requester or customer by delivering as much as possible at a given time and promising the still missing features, functions, or sub-deliverables later. In some environments it may instead be used as a delivery strategy, since it allows for quick wins

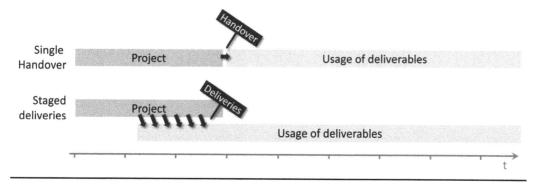

Figure 1-3   The time-relationship of the project and the usage lifecycle of the project deliverables are different in a single delivery and a staged deliveries handover.

by delivering rudimentary products early. Users of these semi-finished products can then be observed and surveyed to identify and fix weaknesses of the product before it is finally finished. Figure 1-3 illustrates the differences between the timing of the project and the deliverable usage between single and multiple deliveries.

While staged deliveries have become a normal practice in many industries, they are rarely described in textbooks, maybe with the exception of project management writings in the field of "Agile" methods. The presence of staged deliveries fundamentally changes cost-benefit calculations because the time that the investment in the project is done overlaps with the time when the first benefits can be created from this investment.

Staged deliveries often come with multiple deadlines. It is often assumed that a project has one deadline, by which it must be finished and its deliverables handed over to the internal requester or customer. In reality, many projects have a sequence of deadlines, and each of them may be linked with penalties or damage claims. Often, these deadlines have been decided upon before the project was initiated and no one checked them for plausibility and feasibility. They may have been part of an "Invitation for Bid" or a Request for Proposal (RFP), and when the business development staff made a decision to respond, this included a commitment to meet the deadlines. Later, when the contract was awarded, a project manager was selected and the deadlines were examined for the first time. Similar situations occur for internal projects. A major requirement for many project managers today is to meet the commitment and manage the project along such a sequence of deadlines, each of which may be difficult to meet.

Multiple deadlines are an example of how projects intersect with operations and with other projects inside complex programs. The deadlines are used as a means to coordinate the projects and force them into a degree of predictability and order that is typical and commonly achievable for operations but hard to manage in the unpredictable environment that surrounds projects.

1.4.4 The Growing Significance of Stakeholder Orientation

A major reason for project failure is lacking or ineffective stakeholder orientation, which can often be observed to lead to delays, budget overruns, and organizational or even political upheaval. Interesting examples are the major international trade agreements that are under development while this book was being written, especially the Transatlantic Trade and Investment Partnership (TTIP). The TTIP is a project performed by the United States and the 28 countries that constitute the European Union. TTIP is intended to create a free market in which over 800 million people live, who make up 60% of the world's GDP (gross domestic product, which is the value of all goods and services produced over a period). The project was officially launched in February 2013, negotiations began in July 2013 and were planned to be finished by the end of 2014*. An essential part of the project was the designation of certain people and organizations as stakeholders, most of them industrial lobbying organizations. A normal process in project management would be to identify the stakeholders of the project; analyze their needs, wants, expectations, and potential influence on the project; and then orient the project based on the results of these analyses to ensure their proper information and engagement. The TTIP team instead defined who it would consider as stakeholders and who it would not. The public, the over 800 million people who would constitute the intended common market, were not among the groups and persons selected as stakeholders. As a consequence, the public discussion was taken over by groups that were highly negative on the project, and while many opponents of the agreement did not have much knowledge of its contents because of the secrecy surrounding them, deep-seated distrust has developed in both the negotiating teams and the mechanisms applied. For a project that needs the unanimous agreement of 29 governments, distrust is a predictable cause for crisis and possibly failure. Interestingly, the second similar agreement is under negotiation between the United States and 11 countries from Latin America and Southeast Asia. It is called Trans-Pacific Partnership (TPP) and it suffers from the same problem. Trying to reject the status of stakeholders—people who are influenced by the project—may finally prove successful, but it may also lead to difficulties when these people build up resistance against the project that the team may find itself unable to overcome. Ignored stakeholders often come back, and they bring friends, press, and lawyers with them[†].

1.4.5 Availability of Resources as a Core Uncertainty

The term "Resource" is a difficult one because there are multiple definitions. Project management originated as an engineering discipline, and for many decades, the most common definition of resources was the one used in engineering: staff, equipment, and materials. This definition is still found in most software products for project management. During the last two decades, the understanding of what project management is has changed, and today it is also seen as a business management discipline. As a consequence, project management has adopted the definition that business scientists and practitioners would use, which is much broader and would also include funding, know-how, time, services, and many more categories. Anyone and anything whose availability is limited but crucial for the project can be a resource.

There is another resource, a key resource that is often overlooked: management attention. This resource is often scarce, and its availability is rarely reliable. Management attention is a

---

[*] While these lines are written in September 2015, the project schedule has been delayed by one year to have the negotiations finished by the end of 2015 and the agreement ratified by all parties in 2017.

[†] These lines are not intended as a contribution to a political discussion but as an assessment of the risks that projects run into when stakeholders are ignored or when project managers assume that they can decide who is a stakeholder and who is not.

key resource because it is a prerequisite to obtaining all the other resources. There is no guarantee that a project team that possesses management attention will also have people, equipment, etc. as required, but if management attention is not sustained, it is quite certain that the other resources will also not be available for the project in sufficient quantities. Ensuring management attention can be a very difficult task. In many organizations, project portfolios, the collections of projects that are run concurrently inside the same governance domain and use the same set of resources, consist of more projects that the organization can manage. Managers are often enthusiastic about the great ideas that they have and wish to implement, or initiate too many concurrent projects to address the many issues that require resolutions; and often, these managers overlook the fact that the large number of projects may be more than what the organization can handle and what they can support. In these cases, the time, energy, and attention of these managers is the scarcest of all resources, and while projects in a portfolio compete for resources anyway, the competition for management attention may be the most difficult one. To make things worse, projects need some kinds of metrics to assess success and failure, or at least some results of stakeholder surveys or a boss who can confirm the perception of success or failure at the top level of the organization. With too many projects at a time, it becomes quite costly to follow up their contributions to the organization's successes and failures and to isolate the positive and negative influences that a certain project has on an organization from the influences of another one. Then, situations can occur where projects compete not only for resources but also for results—when the success of one project impacts the success of another one. An example where I could observe this effect was a project at an insurance company to improve support for free agents. At the same time, a second project to move sales to the Internet made these agents redundant. The first project was intended to give agents confidence in the insurance company and motivate them to invest more time, energy, and money into sales activities, while the second destroyed this confidence and motivated them to move to another company. Another common observation in over-extravagant portfolios is competition for the acclamation for successes, leading to frustration and burn-out on the side of the project managers and their teams when they perceive a deficiency of prioritization and an imbalance between their efforts and the rewards they receive from management. Without the clear perception of gains from the project on the side of management, it is more difficult in turn for the project manager and the team to obtain resources. The dynamics of success and failure act against the project, not in its favor.

Almost all methodological approaches, including the critical path method (CPM), critical chain project management method (CCPM), Scrum, and PRojects IN Controlled Environments (PRINCE2), presume that resources are available when the project needs them. It is methodologically difficult to look at both aspects of project management at the same time—tasks that team members need to perform and the times of availability or absence of these team members—and the focus in most project management methods is on the tasks. This focus makes it easier to develop and teach a methodology but reduces its value for many project managers. In modern organizations, it is far more appropriate to presume that resources are not available when the project needs them, at least not in the quantities and with the aptitudes that the project needs. Project managers and their teams must become active to make resources available, and they should generally not assume that this availability is reliable over the course of the project, but instead need to protect their resource plan from avoidable disruptions from outside the project. How the availability of resources can impact the project more than the critical path is shown in Figure 1-4 through Figure 1-6 (on next pages).

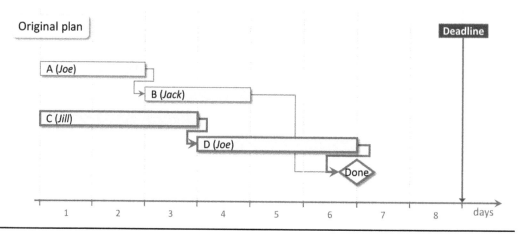

Figure 1-4   A combined barchart-network-diagram (often called "Gantt chart") showing four activities A–D performed by three team members—Jack, Jill, and Joe—and a milestone, "Done". All work must be finished by the end of the eighth day. The path C–D-Done is the critical path, because it defines the earliest date that the milestone "Done" can occur.

Promoters of CPM say that a project manager must keep the focus on the critical path, as delays there will impact a finish date more than delays of critical activities that are not on this path; however, the example shows that seemingly uncritical activities can need more attention. The example further shows that presuming unlimited availability of resources leads to unrealistic plans. In reality, plans with "phantom resources" are a familiar occurrence, with resources over-allocated by numbers assigned during periods of unavailability. Sometimes one can even see resources allocated that need to be booked timely and formally, but no one has done this timely enough. Planning with phantom resources means planning to fail, and while one should not expect that a plan simply works, the part of the plan that leads into failure generally does.

The discussion of the example still ignores other open questions that surround resource planning, such as limitations of availability of resources when activity start and end dates are

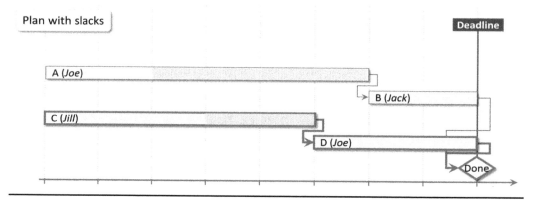

Figure 1-5   This version of the plan shows the maximum delays ("Slacks") that the ends of activities A and C can have without causing a missed deadline. The slack for activity A is four days, for activity D, the slack is two days, so activity D is more critical and needs more attention than activity A.

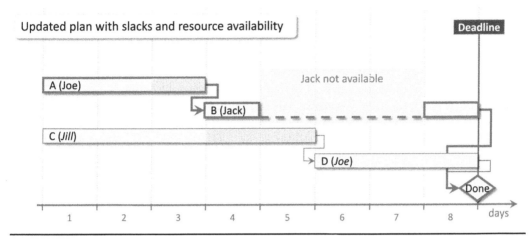

Figure 1-6   More information is added to the plan: Jack will not be available on days five through seven. This reduces the slack of activity A to one day. In the example, Jack would need to begin activity B on the fourth day, then interrupt work for three days and finish the activity on the eighth day. The order of critical activities has moved to A–B–Done, which means that more attention must be given to protect these activities from delays than is given to C and D.

changing. Most project managers have experienced situations in which a short delay in one activity translates into a long delay in another one, because a window of availability is missed and one has to wait for the resource's next available time window. How reliable is the planned resource availability? Is it possible that a committed resource will not be available as promised because of other assignments with higher priority in operations or in another project? Ignoring the limitations and uncertainties of resource availability leads to project schedules that cannot be realistically implemented. A planning approach that considers all three inputs to planning—the flow of work as described in a network diagram, the often unevenly distributed availability time windows of resource availability, and the volatility of commitments—may become too complex for manual planning and tracking. If one decides to develop a robust schedule, one probably needs to use special project management software, as the complexity may overwhelm manual planning.

1.4.6 New Requirements on Procurement in Complex Multi-Tier Supply Networks

The intensity of procurement is increasing. One can easily observe that in make-or-buy decisions, where five or 10 years ago the "make" option would have been chosen, today the option selected is more often "buy". Innovation speed has become too fast for employees in large corporations, such that these organizations have to procure already developed skills from outside. Plus, instead of simple customer/contractor relations that may constitute linear supply chains, we find complex *n*-tier supply networks crossing national borders. Among the many problems that come with them is the lack of project managers who know how to manage these networks for project success. An example of the problems that project managers need to cope with is the Boeing 787 Dreamliner, which ran into problems when most of the development work was done by subcontractors, sub-subcontractors, and so on—companies with whom Boeing did not have a contract at all. At times, no one at Boeing had full knowledge of what companies were part of its project at all. The number was so large and not all were visible to the aircraft manufacturer, as many were vendors of the vendors. One of these subcontractors was a member of the Airbus group, the direct competitor of Boeing, and when Boeing found this out relatively late, suspicion was high at Boeing that this vendor would not do its best to support the development program. In a contractual relationship, parties generally try to protect themselves from damage claims and liability. Most companies have experience that project conflicts sometimes turn into lawsuits, and the parties act in a way that prepares them for a highly competitive litigation. Another aspect that calls for competitive behavior is the open sharing of know-how, which is necessary to make the partners work as a team, but is also considered a loss of organizational assets for the sharing party. While the parties must protect themselves and their assets, to act as a team, they must be organizationally and interpersonally connected and turn to highly cooperative behavior with the intention to meet a common goal: make the complex system work. From a legal view and considering their intellectual property, the parties should surround themselves with walls; from a project perspective, these walls should be ripped apart and be replaced with mutual trust. These conflicting interests and behaviors are hard to reconcile, especially as project managers rarely get trained to do so.

Cooperation needs trust, and for the people at Boeing, trusting their competitor was a difficult task.

To add a further problem, members of international supply networks act in different legal environments. About a third of the design and production work for the Dreamliner has been contracted to Japanese companies*. The United States is part of the Common Law legal system, whereas Japan is a Civil Law country. These two systems have major fundamental differences that even affect the definition of what a contract is. An example of these differences is the amount of documentation that a company under contract produces: In the US Common Law environment, a judge can require a party to hand over all documentation in a lawsuit, and the party should then comply with the request. This is different in most Civil Law jurisdictions, where the parties are free to decide which documentation they use in a lawsuit. In Civil Law environments, it is good practice to document every detail of performance under contract, because this can become the evidence needed to win a lawsuit. In the United States, a party may decide to not create certain documents to avoid their potential use in court. Conflicts on documentation can, at times, become interpersonal conflicts and jeopardize project success. It is a necessary practice to define the applicable law in a contract, and each party commonly tries to make its home jurisdiction the applicable one. In addition to the legal question, there is another aspect which is rarely discussed. At least one of the contract partners will have to work in an unfamiliar jurisdiction. Opinions, behaviors, and rules of business that are considered appropriate and in compliance with law at home may be inacceptable and possibly illegal in this foreign jurisdiction. In a "Mission success first" setting, the parties should consider the misunderstandings and conflicts that can arise from this situation and nurture their relation to avoid any need for legal proceedings.

Cultural dimension can also become a problem when contract partners are located in different countries. Corporations running international projects have developed capable systems to develop cross-cultural competencies; however, proper behavior under foreign law is rarely taught.

---

* See (Gates, 2007).

1.4.7 New Approaches Continue to Emerge

New concepts, new knowledge, and new requirements continue to be developed rapidly, including Agile approaches, leaderless organizations, game theory to explain competitive stakeholder behavior, and online software services to support virtual team work. The situational approach described in this book may be a fourth one. The older tools and techniques of project management can and should continue to be used but may not be sufficient. They are often based on presumptions that were realistic in earlier times and certain situations but are not as useful today. Many get overburdened with expectations that they cannot meet and model the world using simplifications in areas in which understanding and managing unavoidable complexity may be essential for success.

The only constant in project management is change. Some people feel uncomfortable that project management is a discipline that requires life-long learning and self-improvement; others find this requirement among the most interesting and challenging in the discipline.

1.5 How We Are Seen by Others

Most know the following situation quite well. A project manager is driving in the car on a freeway headed for an important meeting, important enough that the person must be there on time, and there is not much time left. The person is driving fast, possibly a bit too fast, then he or she notices a sign for a road construction site ahead and a traffic jam before the site. While the person is slowing down, what will go through his or her mind? Will the person think, "Hey great, they are making best use of tax payers' money" or, "They are removing last winter's pot holes; the freeway will become wider, faster, safer, and more comfortable to drive. I am looking forward to the time when they will be finished, and I can drive here again!"? Rather unlikely, and it is probably appropriate to avoid using the words that most people would use in such a situation. Road construction workers often see people who are angry and are making inappropriate gestures, and some even report that they have seen drivers pointing a gun at them.

As a motorist on a freeway, or any other road, a project manager becomes part of operations just like any other driver; road operations, of course. Supporting vehicles to travel from one place to another is an ongoing and repetitive business. Road constructions are temporary and unique endeavors, or projects. Even the most passionate project managers often forget that the people managing the sites are essentially their colleagues. Road construction may be a Greenfield project, building a new road in a virgin area, and most will love to see it progress and will look forward to the opening. Most road constructions, however, are Brownfield projects; they are largely put up inside existing infrastructure and will disrupt existing road operations. The number of lanes is reduced and the remaining ones are narrowed. Drivers have to reduce speed and may even have to stop from time to time. If a car breaks down or an accident occurs in a road work area, there will be no shoulders and no emergency lane to park the car out of the traffic. The bottleneck character of a construction site on traffic commonly creates jams, and the proximity to big trucks in neighboring lanes may even terrify some drivers. In addition, road constructions are preferred locations for speed traps. A road construction site is a typical example of an operational disruption caused by a project.

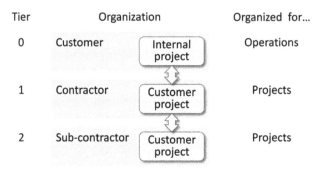

Figure 1-7 A common situation in a project with contractors on various tiers: Contractors in project management may be established and structured for projects, but the customer is not.

Most organizations are not set up for projects since their business focus is operational. This applies to some degree even to project managers on customer projects: Their own organization may have a business focus on projects, but the customer does not. Maybe it is the customer of the customer who is operational, if you work at a subcontractor organization, for example (see Figure 1-7). This may be the location where the subcontractor has to do the majority of the work. Internal projects, as much as those done by a supply network, are perceived by the operational requester or customer as much as disruptions of their business as motorists feel troubled by construction sites.

Most organizations are highly optimized for operations; they are doing the same business over and over again and generate income with ongoing productions and services. At first glance, the repetitive character of their business may not be observable because modern configuration management has made these repetitive undertakings highly adaptive and personalized to the needs of customers, but once one looks at the core activities of these activities, they are basically repetitive. Managers of modern automotive companies often proudly state that their company does not make two identical cars on any given day. While this statement may be true, it does not say that each car is unique; only the configuration of a standard car that is unique. The basic car remains a standard car, a mass product made by numbers.

The practitioners and pundits in project management are rather considered disruptive aliens in this world, comparable to those people working on road constructions on freeways. In these organizations, operational managers bring money home, not project managers, and while operations creates an environment of predictability and order, projects generally bring uncertainty, restrictions, and disruption—this is at least the perception of most operational people. Many project managers regard themselves as heroes, but in the opinion of operations managers, their reputation may be much less valued. A production manager at a major German automotive company described this perception to me some time ago by saying: "If our production people would do as badly in meeting deadlines as our project managers, the company would have run out of business long ago".

This reputation problem may get even worse. Some years ago, I was confronted with a crisis that some of my training students had. Their project needed the support of employees of a customer organization—a major government agency—to implement human resource management (HRM) software in the agency. The project people, employed by the contractor, needed to understand the processes applied by the customer in order to customize the standard software that should be implemented and adapt it to the peculiarities of the agency. By that time, they were facing a lot of resistance inside the organization. Employees did not have to fear losing their jobs, as they were guaranteed that the software implementation would not lead to layoffs. However, their domains of influence would change and, to some degree, possibly their working environment. Some of them would possibly be transferred to other departments or assigned to other tasks, but the employer had ensured them in writing that no one would be demoted or paid less in the new position. Employees were nevertheless afraid of the changes. Many had developed their offices to their own little paradises over the years, decorated with ceiling-high plants and orderly framed children's paintings, and they felt so familiar in their intimate spheres of influence, they perceived any change as a threat to their comfort zone. The more they gave away their knowledge to the project team, they felt, the more they risked losing their homely and cozy environment. Resistance to the project was not open, it consisted of little disruptions here and there, and it was often difficult to impossible to identify the sources of these disruptions. As a result, the project could not continue for a long time, and even managers of the agency were not able to help. They felt that they could not make decisions against their employees, since their success was based on the employees' work. Another frontier of resistance was complaints. The project managers were accused of not being respectful and considerate enough toward the agency's employees and their lifetime achievements. The project finally succeeded, but it took much more time and money than originally expected. Major parts of the project were run under fixed-price contract; therefore, the contractor finished the project at a financial loss and with resources tied up for months that could have been more profitably used in other projects.

The example shows how, on top of the operational perception of disruptions, another reputation marauder may make life difficult for project managers: resistance of stakeholders to change, even if they will not suffer any kind of objective loss.

Project managers are aware of their contribution to organizations and to society in general: Without modern road construction, for example, roads would generally still have one narrow lane in each direction and would be winding around hills and pass through neighborhoods that they quickly bypass today. They would have more holes than a piece of Swiss Emmental cheese, and they would be unable to cope with the numbers of cars using them today. If there are notorious traffic problems at many places, this is rarely due to road construction, but to a lack of it, such as when sufficient public money and political will are not available and are needed to keep road systems in shape and capable of accommodating growing traffic. In the previous example of the government agency, many employees learned at the end that their new offices were much nicer and their new areas of influence were more satisfying than the old ones, and it led to an unexpected bit of additional quality of life. But this could not reduce the financial loss made by the contractor.

Project managers should be aware of their contribution to organizations and society: Without a constant stream of results from research and development projects, without the improvements that our projects bring to organizations, but also without our efforts in demolition and renaturation of facilities that have become obsolete, most of the organizations that rely so heavily on the effectiveness and efficiency of their operations would no longer exist.

1.6 The Complex Dynamics of Success and Failure

San Francisco has three major bridges: the Golden Gate Bridge leading from San Francisco to the North and the Bay Bridge (or, officially: the San Francisco-Oakland Bay Bridge), which is actually two bridges to the East: The first stretches from the city to Yerba Buena Island and the second from the island farther on to Oakland. The three bridges were all built between 1933 and 1937 and have a similar length (2,700 meters or 9,000 feet for the Golden Gate Bridge, 3,000 meters or 10,000 feet for each section of the Bay Bridge). They also had a similar price tag ($35 million for the Golden Gate Bridge, $77 million for the two Bay Bridge sections combined). In September 2013, a modern replacement for the Eastern section of the Bay Bridge was finished, and traffic was diverted to the new bridge. The replacement was necessary because of provisionally repaired damages by a 1989 earthquake and because of fears that a much stronger future earthquake might completely destroy the bridge. New construction was chosen as an economically more viable solution than reinforcing the bridge and substituting weak components with stronger ones, and the old bridge was simply considered "ugly" by too many people.

It would be an unthinkable sacrilege to replace the Golden Gate Bridge. It is regarded as an engineering icon of the United States and was labeled as the most frequently photographed bridge in the world. The American Society of Civil Engineers listed it among the Seven Wonders of the Modern World[*], and 10 million visitors every year will probably agree with this statement. With only minor damages, it resisted the earthquake of 1989, the one that had heavily damaged the Bay Bridge, but in order to make it also withstand a more disastrous earthquake, a retrofitting project was initiated in 1997, which was expected to take more than 20 years to finish. Care was taken that the shape and beauty of the bridge were not impacted by the reinforcements. This retrofitting project was a clear case of a project success in the classical architectural values of durability, convenience, and beauty[†]; in addition, it was finished under (final) budget and ahead of schedule. It is hard to deny that the bridge project was a success, something one may call a "Wow project".

There is one exception, however. Joseph Baerman Strauss, the project manager of the bridge, had a project objective which was innovative at that time: "Zero workers killed". The rule in the 1930s for high steel construction projects was one fatality for each million dollars spent. Strauss made sure that site workers wore hard hats and followed the safety regulations he imposed. In addition, he strung a safety net under the construction site, an invention that saved 19 workers' lives; those workers later established a club called "Halfway to Hell". In two accidents, however, the net did not help: One worker was killed by a falling derrick arm, and 10 were killed when a travelling platform, on which they were working, fell down and ripped the net. A death toll of 11 people in a project worth $35 million was a very good record for the time of the bridge's construction, but the project objective of zero fatalities during construction was not achieved.

Success and failure in projects seem easy to judge at first glance, but they are not. The problems begin with the definition of what makes a project successful and what is regarded as a failure. Different people may have diverse views, and over time, their opinions may change. There are of course projects that are just pure failures, sometimes called Zombie projects. They

---

[*] (ASCE—American Society of Civil Engineers, 1996).
[†] The Vitruvian virtues of Utilitas, Firmitas, Venustas (Vitruvius, ~15 BC).

should never have been initiated because they had no chance to succeed from the start. Many other projects run into disaster unnecessarily; they would deliver their results if they were managed properly and if they had received the support and attention from the key stakeholders that they needed. Some projects fail because of poor project management or are impeded by team members with insufficient technical, organizational, or interpersonal skills. Projects may fail due to over-organization or under-organization. No doubt, some projects are 100% failures. Then, there are projects that are mixtures of successes and failures, which probably constitute a majority of projects. If the perception of success is stronger than the judgment of failure, one may then call the project successful. A difficulty here may be that different stakeholders may have different perceptions of success and failure. A close relative of one of the 11 dead workers of the Golden Gate Bridge project may have considered success and failure of the project differently from a person who just happily attended the opening celebrations or from the commuters who used it daily as part of their normal way to work.

Can there be projects that are nothing but successes? They are as unlikely as sports teams that win all their matches during a league season; even the best team will normally play a tie game or lose a match from time to time. Perfect seasons do exist. The Miami Dolphins had a "perfect season" in 1972. They finished 17-0-0, but this was unique in 95 years of US National Football League (NFL) history. A perfect project is probably as rare as that season. Success in this understanding is not the absence of failure, but the prevalence of successes over failures in the perception and judgment of a majority of relevant stakeholders. Projects without at least a grain of failure may not exist at all.

It may be a good idea to accept a certain degree of errors and failures as normal in project management, as long as no human life or other similarly valuable things are in danger, and as long as the center of gravity is on the success side. Such an approach reduces the pressure on the teams and allows stakeholders involved to learn positive lessons from them.

1.7 Standardization and Certification in Project Management

Standardization is a generally difficult topic in project management. An essential aspect of projects is their uniqueness. It is common in projects that teams do work that they have not done before and may possibly never do again when the task is finished. This aspect limits the portion of project management that is open for standardization. On the other hand, standards can help project managers and their working environments to better understand their tasks, be better prepared for recurring problems, and be able to avoid expensive misunderstandings. There are generally two types of standards in use for project management: prescriptive and descriptive. There are several prescriptive standards, one may also call them "cooking recipes", for project management, such as:

- PRINCE2 from the United Kingdom
- V-Modell XT, a standard mostly used for public projects in Germany
- Hermes, which serves a similar purpose in Switzerland

Agile methods such as Scrum, the Crystal family of methods, or Six Sigma (a group of project management methods used in quality management), are prescriptive standards for project management. They allow for some freedom in the specific application, just like a cook who has run out of a certain ingredient may replace it with another, similar ingredient, but would still be considered in general observance with the original recipe. Prescriptive standards tell project managers and their teams what they have to do and in what order. This contrasts with non-prescriptive, but descriptive standards such as the *Project Management Body of Knowledge (PMBOK®) Guide,* by PMI®, the Project Management Institute; the British BS 6079 series of standards; and the German DIN 69901 series of standards. There is also a global ISO 21500 standard, which builds bridges between these three standards. To stay in the metaphor, descriptive standards are not cooking recipes, but instead describe what a professional project chef should be competent with and explain what equipment is needed or will at least be helpful in the project kitchen.

Why would a project manager want a standard for his or her work? The benefit is unification of terminology. (The traps that come with confused terminology are discussed in the following section.) Another benefit is that one has a widely accepted reference. If the project manager considers it appropriate to have a project charter*, a written mandate for the project, there is a reference that describes what it is, which makes it easier to insist on such a document and to ensure that different people mean the same document or practice.

The most comprehensive and widely used of these standards is the *PMBOK® Guide*†, and as a project manager, one should become familiar with it. The *PMBOK® Guide* gives a project manager a degree of firmness in terminology and eases communications and decision making. One should also consider getting certified by an organization or company that is associated with the standardization, such as the *PMBOK® Guide.* Project managers sometimes state that they get treated with more respect in their organization since they obtained certification, but this does not actually make them better project managers. Project managers should consider learning a lifelong and continuous process, and the preparation for a certification can be a valuable part of this process, which should not finish once the exam has been passed. As a baseline, knowing a standard and being certified can be helpful, but one should not stop at this point.

1.8 Terminology Traps

The term "stakeholder" is a good example of a word that is commonly used but often leads to confusion, because people use different interpretations of who the stakeholders are in a project. I actually observed four different definitions:

1st Definition—Owners of Tangible Investments

This definition probably goes back to the late 1950s of the 20th century and is easy to understand if you try to describe a project as a kind of company—often within a company—with

---

* This term is used in the *PMBOK® Guide* and in ISO 21500. BS 6079 calls it "project brief". DIN 69901 calls it "project order" (Projektauftrag).
† *PMBOK® Guide: The Guide to the Project Management Body of Knowledge* (PMI, 2013).

similar purpose and organizational setup. Projects and corporations—following this explanatory approach—have many analogous elements, things that may have different terms, but are nevertheless fundamental commonalities:

- A company is organized for some generic tasks that other organizations also do (think of book-keeping), but also for specific sets of tasks, which set the company distinctively apart from most, if not all, other companies. The same is true for projects, which must achieve specific objectives but also need to act in corporate or public environments with standardized rules, processes, and governance structures, and both types of organizations struggle with meeting both kinds of requirements given their limited resources.
- Companies use descriptions of processes and workflows in the form of texts and diagrams. For projects, various network diagramming methods were developed that essentially serve the same purpose.
- A company is run by one or more managers; a project also is run by one or more project managers.
- A company has employees; a project has team members.
- A company has suppliers; projects have contractors.
- A company has shareholders . . .

. . . and this brought the problem to identify a parallel construction for investors in projects. One can easily identify an equivalent to shareholders by looking at gambling: stakeholders, gamesters, who put their money or other fortune at stake. If their game is successful, they gain a benefit from the stake. If they lose, their money is lost. Gamblers play the game if their hope for gaining is higher than their fear of a losing their stake money[*]. This understanding can be found in an older interpretation of who the stakeholders are, such as people who make a tangible investment into a project; an investor, a paying customer, an internal project sponsor, and other commonly high-ranked people who invest money, people, equipment, facilities, and other hard assets[†] into a project. Following this definition, the number of stakeholders in a project is rather small: supervisors, one or more customers, and other supportive managers. It may well be that a project has only one stakeholder, a customer, requester or, using a term from the "Agile" world of project management, a "Product Owner". (Agile methods will be discussed later.)

An example of this definition can be found in an interview with two experts in a publication in PMI's member magazine, *PM Network*[‡], which discusses the importance of stakeholders' early agreement on requirements, and when they say stakeholders, they mean senior management.

---

[*] To add a little reality check on this expectation: Making a big number of people lose their money during gambling is the primary business model of any casino or game hall. If one simply looks at the big facilities that these corporations hold in places such as Las Vegas, Macao, Monte Carlo, and others, and considers both the investment and the running costs that they incur, losses of gamblers must be a very effective source of income.

[†] Please see the Glossary for the definition of terms used in this book.

[‡] (Boyle & Mayes, 2012).

2nd Definition—Owners of Tangible and Intangible Investments

Investments into projects do not need to be tangible. Contributing team members also do investments in the form of soft assets: skills and talents, energy, motivation, inspiration, creativity, discipline and order, the preparation to step into conflicts and the risk of final failure—there are true "Zombie" projects launched by organizations every now and then. They are uncomfortably frequent, and can ruin project managers' resumes and self-confidence, especially those of young project managers, who are often assigned to projects in which they have no chance to ever meet requirements but will have to take the full blame for the predictable fiasco. The preparation to take such risks is also a major intangible investment by a person, and a person may reject the investment if the perception is that there is nothing to gain.

Promoters of this definition consider these intangible investments as much an element of the dynamics of success and failure of projects as the tangible ones, and the basic rule of investment works here as well. People will make the investment if their hope for a benefit is higher than their fear of losing their stake.

Along this definition, the group of stakeholders is growing larger. Here, it includes the stakeholders from the first definition plus all active contributors to the project, taking into account the project manager, the project management team, and finally the entire project team including contractors.

An example of a document which uses this definition is the British Standard BS 6079-2:2000[*], which defines the term stakeholder as "a person or group of people who have a vested interest in the success of an organization and the environment in which the organization operates". One should add that a project organization, following the same standard (p. 10), is a "structure that is created or evolves to serve the project and its participants". In this understanding, project stakeholders are people or groups who have a vested interest in the project and the organization running it based on the tangible or intangible contributions they make.

3rd Definition—Persons, Groups, and Organizations that Potentially Interact with the Project

The term "Stakeholder" did actually not originate in gambling but in gold-mining, from which gamblers borrowed it later. During the times of the great gold rush, prospectors needed to have a formally assigned claim—a mining ground—registered and paid for before they could start panning and digging, and the corners of the ground were commonly marked with wooden poles, called stakes. The stakeholder in this understanding is the owner of the ground marked by such stakes. Based on this metaphor, a project stakeholder may be anyone who has claims that the project needs to consider. In this broad understanding, the number of stakeholders can be millions worldwide. The definition includes the first two definitions and adds all those who are not actively participating but are potentially influenced by the project and who in turn may influence the project. Their interests may not be vested at all—at least not in the opinion of the promoters of the project and those contributing to it—but project managers and their teams need to take into account their potential support but also possibly their resistance. In some cases, this understanding may even include competitors.

---

[*] (BSI, 2000, p. 12).

An example of this interpretation of the word stakeholders is the *PMBOK® Guide,* which since its first edition in 1996[*] emphasized that stakeholders are all people, groups, and organizations who may be influenced by the project and also have the potential to influence the project, positively and negatively. This definition has remained valid in the all following editions until the 5th Edition[†]. Following this definition stakeholders are all those individuals, groups, and organizations who need to be identified and their interests, fears, wishes, expectations, etc. considered during the lifecycle of a project.

4th Definition—Stakeholders by Selection

This definition is dangerous because it has led a number of projects into crisis, commonly on large projects. According to this definition, the project manager, together with some high-ranked sponsors, supervisors, and possibly politicians, make a decision concerning who they consider stakeholders and whose influence they are therefore going to accept and whose they will ignore or reject. It may well be that they keep the project secret from the non-stakeholders as long as possible. The problem: rejected stakeholders should not be expected to remain quiet; they will come back later, requesting changes or termination at a time when it is possibly much more detrimental to the project, and often enough, they build alliances and bring their friends with them. An example is the Denver, Colorado, airport between 1992 and 1995, when airlines, who were rejected from the decision processes, realized that a baggage handling system was under construction that would not be able to handle skis in a new airport in the heart of a winter sport region[‡]. The Transatlantic Trade Partnership between the US and the EU countries, mentioned earlier, suffered from the same misunderstanding and faced the same problems when citizens started asking what politicians and government agents were negotiating behind closed doors, assuming that the results of these talks might violate their interests.

Taking the term "stakeholder" as an example, it is easy to see how much care project managers should take with language. Many have had experiences as to how misunderstandings can damage a project, and unclear language is a cause for many of them. Often, misunderstandings come from terminology that is understood differently, an effect that is often explainable by history. Adding to the problem, dictating the glossary generally serves as a sign of authority, more on a de-facto level than on a formal one, and hidden behind discussions on the correct use of a term are often power struggles among, well, stakeholders. A further cause for traps in terminology are older definitions, which are still in use by some experts and practitioners, while other people have turned to definitions that have been developed more recently or have emerged and then have been accepted in our discipline. This book lists some terms and acronyms that are inconsistently used in project management in the Traps in Terminology section, at the end of this book appendix.

Are we from project management alone with the risk of misunderstandings that are due to terminology traps? Probably not. Even highly structured and standardized schools of thinking, such as the natural sciences, have misunderstandings on fundamental issues. An example is the distinction between metals and non-metals in chemistry and astrophysics. In chemistry, a line

---

[*] (PMI, 1996, pp. 15–17)

[†] (PMI, 2013, pp. 30, 563)

[‡] (Calleam Consulting Ltd, 2008, p. 6).

can be drawn in the periodic table from under Hydrogen over Boron to Astatine, and elements to the right and top of the line are regarded as non-metals, while those to the left and bottom of the line are metals. Seven elements on the line are so-called metalloids, elements in a transition status, between metals and non-metals. Astro-physicians have a much simpler definition: they would instead draw a separating line from under Hydrogen to under Helium; these two elements would be regarded as non-metals, while all others are metals. Figure 1-8 shows the difference.

Many projects are cross-functional and span various disciplines, each with its own terminology and often with terms, possibly entire phrases, used differently. Project managers have to ensure that these differences do not cause misunderstandings that are often difficult to identify and costly in their consequences. In addition, they must also take care how we are handling terminology traps inside our own discipline mentioned above. Often, experts and practitioners are not only unaware of the different meanings of project management terms but reject any discussion of them. They signal this by beginning sentences with, "Everyone knows that . . ." or a similar statement. As a trainer in project management classes and when writing articles and other text documents such as this book, I rather assume that different people do not have the same understanding of the things that I am talking or writing about and strive for clarification first.

For readers of books, articles, and other sources, where one cannot simply ask an author for his or her interpretation of a specific term, a three-step process is recommendable:

1. Inspect the text to see whether the differences in interpretation are relevant for correct understanding. If they are not relevant, simply ignore them. If they are relevant, move to step two.

2. Try to find out from the text what the interpretation is that the author is using. In the case of the interview in "PM Network" previously mentioned, it was easy to identify from the statements of the persons in the discussion that they use the most traditional definition of the term stakeholder—i.e., investors, senior management, and customer. If the interpretation cannot be derived from the text but is relevant for understanding the text, move on to the third step.

3. Make an assumption. It is important to not forget that it is just an assumption and may be proven false later, so be prepared to possibly read the text a second time and then with better knowledge.

This book is about situational intelligence in project management, and a careful or even doubtful use of assumptions and interpretations is a central element of any situational approach. Doubt is among the most powerful tools that a project manager has, and project managers should never give this tool away. In real life project practice, which is mostly a business practice, we often meet seemingly doubt-free people, and many of them can be impressive in their self-assuredness and security. Often, they can easily be identified by an inflationary use of terms such as "best practice" and "excellence", and they always know whom to blame for troubles and failures. Give these people a complex project to run, and they will soon discover the limitations of their approach.

For this book, I try to be as unambiguous as possible in language and provide you with a clear understanding of the definitions of difficult terms when they first occur. In the list of terminology traps at the end of the book, definitions we use are marked in bold, so if you are familiar with a different definition for a certain term or have a different understanding of a concept, you will not misunderstand our statements.

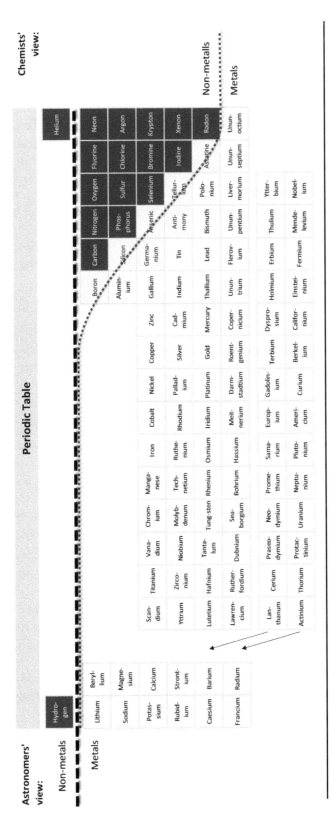

Figure 1-8 An example of cross-discipline terminology traps: In the periodic table, chemists and astronomers have a different definition of metals and non-metals.

1.9 Navigating between Monsters

Antique mythologies are a treasure full of pearls of wisdom that can help project managers improve their understanding of situations they are facing and develop situational approaches to mastering them. An example can be found in Homer's *Odyssey,* where Odysseus, the hero, king of Ithaca Island, has to pass a narrow strait between two monsters called Scylla and Charybdis. Scylla is a flying monster with six heads that threatens to capture half a dozen of his men, one for each head, and eat them alive. On the other side of the strait is Charybdis, a maelstrom monster, which swallows huge amounts of water from the ocean three times a day and then spits it back into the sea. Avoiding one monster would bring Odysseus's ship too near to the other. His crew had to row the ship along the strait between the monsters. But he followed the advice to keep this ship nearer to the side of Scylla than to that of Charybdis because Scylla threatened to take only six men while Charybdis would destroy the entire boat with all men. While the ship passed the strait, the crew had to row the ship as fast as they could, to not become victims of both monsters.

Similar maneuvers are common in project management. The stakes may not be as high as for old Odysseus, but the principle applies:

- Project managers must often choose between heroism and collaboration. There are tasks that are best done without too much consideration and possibly alone, getting things done while others are still discussing the objections. Other tasks require interpersonal and organizational skills, and active and patient listening may then be more favorable than getting things done quickly.

- Some projects necessitate long-term planning and booking of resources on a granular level. Others may suffer from the inflexibility and lack of adaptability and agility that comes with long planning horizons and high granularity. The project manager and the team have invested a lot of time and energy into the plan, and every approved change request becomes a major challenge for re-planning and re-booking. Both over- and under-planning are true monsters, and project managers have to find their way between them, and as in the myth, it may from time to time be better to do more planning or less.

- When project teams develop new products or services, they are often tempted to run fast and overlook important issues. Then they may not have enough time to thoroughly test their deliverables and fix errors found. In their high-speed approach they may also con-sider 95% finished deliverables completed, ignoring that the remaining five percent may drive the project and its stakeholders into crisis. Other teams may be highly diligent and drive their deliverables to absolute perfection, but miss deadlines with this approach or launch the product or service at a time when a market window has closed and customers are no longer interested. Too fast and too slow are both monsters between which most projects must navigate.

- A project manager may decide that for a given task a highly expensive solution is the most appropriate one, and that the investment in this solution will pay itself back during the project or after it has ended. If the project manager does this too often, the project may at one time run over budget and in the worst case may need to be terminated. Another proj-ect manager may instead choose the least expensive solutions available, thus protecting the project from overruns. This approach may lead to results that the customer or an internal requester will not accept, or to problems that may turn up much later and diminish the success of the project for its stakeholders. Project managers have to make many decisions considering the two risks of being "too expensive" and "not good enough".

- Project managers should always stand for realism. Realism is a place between two extremes: easy and panic. Both standpoints have their justification in organizations, and most companies have experts for them. Salespeople are mostly optimists, and it is part of their profession to transfer the expectation of easiness and comfort to customers and also inside the organization. On the panic side are experts in worker protection, data security, quality management, fire prevention, and more. They must be perfect pessimists and picture worst-case scenarios for everything they assess. Project managers should place themselves between the two extremes. They should build their plans based on expectations and assumptions that they consider likely and realistic, being aware that things may be found later that are easier or more difficult, inexpensive or more expensive, faster or slower, and so on.

- Project managers are forced to find a balance between the two types of action: proaction and reaction. Proactive approaches are implemented through forecasts, plans, commitments, and agreements. Reactive approaches are characterized by responses to situations, improvisation, workarounds, error fixing, and decisions in states of emergencies. Proaction can reduce the costs of projects, improve the availability of qualified resources, and gain trust by stakeholders. It can also collide with the need for agility in changing environments and with the appropriate management of unpredicted events and conditions.

- While project managers are commonly expected to develop a performing chain of direction, they often have to report to their team and to lead their sponsors and other supervisors. This may sound counter-intuitive, but even project managers who have a good understanding of technical matters or other special knowledge in the application area of the project must rely on the skills of their team members to ensure informed decision making. A typical aspect of complex projects is their cross-functional, cross-organizational, and even cross-industrial setup, and projects stretching over countries and cultures are also common. Decisions made by individuals who are most knowledgeable in the various domains may trump traditional hierarchical command and control structures. It may, in turn, also be necessary for a project manager to teach the sponsor and other high-level managers, as they may lack an in-depth understanding of project management. Given the many egos that need to be considered, reconciliation of top-down and bottom-up decision chains are hard to achieve, but this ability may be a key factor for project success.

- Another fundamental conflict is whether we manage by emotions or by reason. Emotions include fear, anger, and moods. A statement in a discussion, "You are so emotional", is rarely meant to be flattering. Emotions can also include passion, team spirit, loyalty, and other sentiments that are generally supportive to projects. Emotions are inevitable to projects, as the brain processes emotions first, and reason second. We often notice that angry or disappointed people are unable to develop empathy for the point of view of others. Empathy is probably the highest ability of human reason, but when emotions block higher brain functions, reason may not guide human behavior. Project managers have to balance both emotion and reason and try to steer them in a way that is protective for both the project and its stakeholders*.

Navigating between two monsters is a difficult task, especially given the many stakeholders around, who paint the threats of one monster in vivid colors while ignoring the other one. Project management includes much balancing between extremes, and the situationally perfect balance may be different from project to project and even from one project situation to another during the course of the same project; therefore, it is difficult to define best practices that can be expected to be suitable for all of them. I will discuss best practices in the following chapter in more detail and provide recommendations for a link of favorable practices with project types. Coping with the dynamic nature of projects requires a high degree of situational intelligence and adaptiveness from all participants, but mostly from the project manager.

---

* When I am leading teams, I generally have a rule that praise is communicated in public, but criticism in private, to broaden the emotions that drive the project and to limit the effect of those that restrain it.

Chapter 2

Digging Deeper

Many practitioners and experts in project management call for best practice solutions and strive for a kind of universally applicable excellence. During more than three decades in project management as a practitioner, a trainer, and occasionally a consultant, I have developed a deep distrust in these concepts, based on personal experience and observations from a distance. The same practices that have been applied successfully in the past may lead to failure in another situation and practices that have not worked in the past became successful in new situations. This chapter lays the foundation on which most of our projects are performed, before Chapter 3 details project topology.

2.1 Introductory Questions

1. You have been assigned as the project manager to a project that involves new technology applied in a complex organizational environment. You find that the actual application of your project management plan does not lead toward project success and that people deviate from this plan, but you are not quite sure if it is the wrong application of a good plan or if the plan is too faulty. Time is running out. What should you do next?

   a) Apply corrective action by ensuring that all team members adhere to the project management plan. Then observe the outcomes of their work and make the decision whether you need to re-plan the project.

   b) Accept that team members do not adhere to the plan and find ad hoc solutions to solve the most urgent problems.

   c) Get earnest feedback from stakeholders, including descriptions of their organizational and technical problems and reasons why execution of the project has not adhered to the plan. Then decide on the degree of corrective action and re-planning needed.

   d) Re-plan your project immediately.

2. A project management team has taken over the responsibility for a new project and begins planning by putting together some first rough estimates on time, effort, and cost as well as on technical and functional details. What should be done during this early planning phase?

   a) The estimates should be documented in a way that allows them to be adjusted and refined later in the project.
   b) The estimates should be baselined immediately to ensure that they are not altered without formal authorization.
   c) The estimates are preliminary and should not be taken too seriously. There is no need for documentation yet.
   d) Every time one of the estimates gets refined during further planning, a change request will be needed.

3. Projects performed under contract are different from internal projects in all of the following aspects except which of the following?

   a) Project managers running customer projects are often in a strong position inside the performing organization because they are responsible for providing the organization's revenues and have to avoid breach of contract.
   b) Resolving disputes in customer projects may require legal action at a court. Disputes in internal projects are resolved inside the organization.
   c) It is a specific characteristic of projects performed under contract that the progressive elaboration of product requirements needs to be meticulously coordinated with proper contractual project scope definition.
   d) Satisfying the stakeholders' needs and expectations is more imperative for project management teams in customer projects than in internal projects run by an organization for own purposes.

4. Which of the following are strong signs that a Nash Equilibrium is about to occur in your project?

   a) Your team members cooperate in a way that enables them to find and implement great resolutions for difficult problems in a short time.
   b) Your team members communicate just as much as is formally required and avoid taking personal risks in the project.
   c) The working style of your team is characterized by the desire to understand the issues of the other team members and by mutual support.
   d) The team members place their common interest of running a great project over their individual interests of short-term gains.

5. You are running a project for an "internal customer" inside your organization. What does this actually mean for your project?

   a) Internal customers and internal vendors are a means to model profit center relations in an internal charging system.
   b) Working for an internal customer, your project provides direct revenue to the performing organization.
   c) Charges by your project to the internal customer are real costs for the performing organization.
   d) You are running a pretty normal customer project.

6. Your project was initiated, and a first rough plan was developed, based on a set of technical and organizational assumptions that later turned out to be false and on estimates that were found to be overly optimistic. This has led to budget overruns and delays, and has caused your team members to work on the project beyond the original scheduled times. During the further course of the project, more overruns are to be expected. Which statement is probably most helpful in this situation?

a) Project managers should not worry too much about costs, schedule, and workloads on their team members; the only success metric for a project is the delivery of the expected benefits.

b) A company performing a project should build a maximum of reserves into their schedules and budgets to cover problems as described here.

c) One should never start a project based on assumptions and estimates, but instead always rely on proven facts when fundamental decisions need to be made.

d) One has to take certain risks to get projects started, and basing decisions on assumptions is one of them.

2.2 A Major Distinction

Chapter 3 covers project typology that emerged in research, but is also based on personal experiences and observations. The type of project or project situation should influence the practices (approaches, tools, techniques, and behaviors) that a project manager applies to run the project. I will try to give some indication of what practice is favorable or detrimental in certain project types in this chapter.

One project distinction, however, is obvious, and its implications are fundamental enough to discuss it here upfront: the distinction between internal projects and customer projects. This differentiation is mentioned from time to time in literature, but we have not yet seen it discussed and elaborated in detail anywhere. The most basic difference is that project managers in internal projects administer cost centers, whereas the majority of their colleagues in customer projects manage profit centers in the truest meaning of the expression. In internal projects, costs occur during the course of the project, but benefits are expected to emerge after handovers, and most of the benefits, or even all of them, after the end of the project. The benefits for the contractor from a customer project come in the form of payments from the customer. It depends on the contract terms and how early payments are agreed upon and expected against the progress of the project, but they should not be too distant in the future, as the liquidity of many project vendors would not allow for them to lay out too much money. It is not uncommon for the customer to pay partially or fully in advance to ensure timely funding for the project that the contractor may not be able to complete without these funds on his or her own. In simple words: project managers in customer projects bring money home.

An internal project may be undertaken to develop a new product or service that can be launched on the marketplace as soon as it is considered sufficiently complete and mature. An internal project may also strive to reduce operational costs, and its business case calculation will consider the savings as if they were revenues. These monetary benefits for the organization will nevertheless not be earned by the project but by the business unit for which the project occurs. Many internal projects are performed without monetary considerations at all; they have to improve the organization internally or help it position itself better on the various markets.

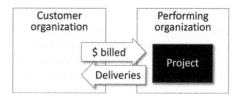

Figure 2-1 Billing (= invoicing) and internal charging are easy to differentiate. The internal charges model flow of money is inside an organization; billing is linked with actual income and cost: Money crosses organizational borders.

Other internal projects have no business case and no financial, organizational or strategic goal at all but must help the organization meet legal or contractual requirements. In all these cases, internal projects do not create an inflow of money into the organization but generate long-term benefits or compliance for organizational units outside the project; at least this is the expectation of the performing organization.

There are situations when the distinction seems to get blurred: Organizations define "internal customers", who may simply be other departments, possibly located three doors down the corridor. Sometimes, the "internal vendor" and the "internal customer" even have an agreement called an "internal contract". Similar to external customers and vendors who have a flow of goods and services to the customer and a flow of billed money back, internal customers get deliveries from internal vendors and often pay for them implementing an internal charging system. Often in such organizations, one hears statements that internal charges are costs just as external, invoiced costs are; but they are not, as Figure 2-1 shows.

In some other projects environments, when there may be a group of companies, subsidiaries under the control umbrella of a holding corporation, and one of these companies works as a contractor for another one, things can be more complicated. A common example can be found in corporations with internal IT providers and consultancies that manage projects exclusively for other member companies of the same groups to which they belong. These companies are registered as their own legal entities, and their projects formally are run under contracts; so their projects are generally considered customer projects too, but what are they really?

The core principle of a contract and the litmus test to separate it from other types of agreements is the legally binding nature of the contract: it is enforceable at court. If a customer project performed under contract is deeply troubled and in crisis, either party may seek remedy by taking legal action against the other one. Problems in internal projects, in contrast, are commonly remedied at management level; often, they are actually swept under the carpet of managers who prefer that competitors, customers, shareholders, and other stakeholders are not made aware of the failures that have happened under their supervision. Departments inside an organization cannot sue each other in court. Sometimes, departments even wage personal wars against other departments, possibly really fierce wars, but they cannot sue each other, because they are not legal entities. Then, the term "internal contract" for such settings is inaccurate, as an internal agreement cannot be enforced by legal action; it can only be escalated to higher management levels for resolution.

What about the companies under a corporate umbrella running projects for customers that are also member companies of the same corporate group of companies? They can have formal contracts with their customers, of course, but normally, a holding company will not allow them to sue each other; this would foil the strategic supremacy of the holding company over its subsidiaries. From their own company perspective, these companies will have a customer project. From a corporate view, these projects will be considered internal, because they are run under the umbrella of the corporate group. As one would expect, the group view is the dominant one; you should regard these projects as *de facto* internal projects. This may change when the holding corporation decides to sell one of the member companies to a new owner, who is not part of the group, but unless this happens, the projects are factually internal. Figure 2-2 explains such a situation.

Another factor that may add to confusion is the existence of customer-side project sponsors in customer projects that are managed by contractor-side project managers. The project sponsor is the person, who mandates the project manager to run a project inside the organization. If the contractor runs the project for the customer, how can the project sponsor be located on the customer side? The explanation is straightforward: in projects performed under contract, there are actually at least two projects: a customer project on the side of the contractor and an internal project on the customer side. Each of the two projects may have its own project manager and project sponsor. The document in which the mandate for a project manager is explained is called the project charter. It is the foundation document for the project and the authorization for the project manager to use organizational resources for the project[*]. See Figure 2-3 for an illustration of the difference between internal projects and customer projects.

The last confusing element for the discussion here is the label "customer project" by external project managers in internal projects. These are still internal projects, as the performing organization, the organization that provides and manages the resources for the project, is identical to

---

[*] In a small company with only a few employees, who are directly lead by their owner/manager, the document may be dispensable and a verbal mandate may be sufficient. In a large corporation, a clear mandate and authorization to spend organizational resources may be necessary to perform any project.

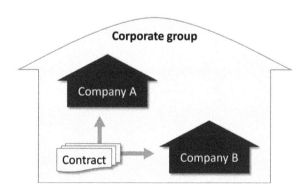

Figure 2-2 Companies A and B belong to the same group. A project that one of the companies runs for the other one may be considered a customer project, and formally, it is done under contract. From a corporate perspective, it will nevertheless be seen as an internal project: in case of a conflict, neither company will be able to sue the other.

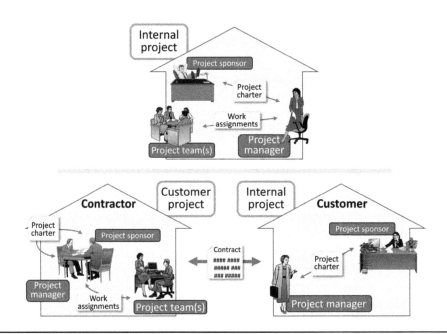

Figure 2-3   In an internally run project, the resources for the project are provided and managed by the organization that will use the deliverables. In a customer project, it is done by another organization working under contract. The customer may nevertheless have an internal project that has the work outsourced.

the organization that will use the project's deliverables. The project manager is just among the resources that the organization procures from external sources for its internal project.

What are the differences for project managers and their teams between running internal projects and customer projects, cost centers and profit centers? There are many, as follows, and they are significant.

2.2.1 Internal Projects

Project managers in internal projects administer investments of the performing organization. These investments are cost centers; sometimes, they may have some quick wins that they can achieve early, but it is generally expected that benefits will come over time, mostly after the end of the project, possibly over years. Benefits may be strategic, monetary, or some other kind of value that the organization expects from the project. Some projects do not provide benefits to the organization at all; they are mandated by law, and the project is done to ensure compliance with this law.

A project manager in an internal project has the fundamental advantage that the person has to consider the interests of only one organization. Another advantage is that the project is performed inside the organization that the person knows well, unless the person has been just hired for the project, and that the person is familiar with the strategic setup, the corporate culture, the market environment, and the key stakeholders of the organization. This advantage comes with several disadvantages. The project has to compete for resources with the lines of the functional organization (and possibly other projects) and is mostly in a weak position when decisions on prioritization need to be made. Project managers in internal projects tend not to have a budget; their responsibility may be limited to technical and functional aspects, to scheduling, and sometimes to certain risk management activities. The budget in these projects is mostly managed by the project sponsor, and the process assets, such as procedures, forms, templates etc. are provided by a Project Management Office (PMO), a governance structure to unify approaches, methodology, and terminology in an organizational project portfolio. These process assets often lack flexibility and limit decision options for project managers. It is also much more likely that the project managers are expected to run the project on a part-time basis. With all these conditions, it is rather unlikely to meet objectives regarding functionality, schedule, cost, and limited operational disruptions.

I observed that these project managers generally have a much worse reputation among colleagues inside their own organization compared to those in customer projects. I already mentioned the automotive production manager, who said to me some time ago, "If our production people were to do as badly in meeting deadlines as our project managers, our company would have run out of business long ago". While this comparison (the predictable world of operations with its far more reliable availability of resources versus the more adventurous world of project management) is generally somewhat unfair, it gives a strong indication of the low status that may often be assigned to project managers.

Another aspect of internal projects is that most are performed inside the protective environment of the organization that allows it to keep any failures a secret.

2.2.2 Customer Projects

Project managers in customer projects provide income to the performing organization. In this role, they become untouchable—to some degree. The projects are run by their organizations under contract for one or more paying customers. In the following description, I assume that there is just one customer, but in practice, there may be several. I prefer the term "Contractor" for the performing organization, but many other terms are used, such as suppliers, providers, vendors, sellers, and so on. Project contractors supply products, provide services, and also have a management function to ensure that their contractual obligations are met in an effective and efficient way. Both organizations, customer and contractor, commonly provide governance functions to ensure alignment and compliance of the project with their strategies and policies and with any mandatory standards, regulations, and laws. So, the project manager must consider and meet the requirements of two organizations, and it is possible that these requirements are contradictory from time to time.

Project managers in customer projects run profit centers and produce an income stream for the performing organization, possibly the only one that the organization has. This gives them some major advantages: They commonly have their own project budget, which is easily defined. For example, in a fixed-price project, the price that has been agreed with the customer minus the margin or raw-profit that the contractor organization is intending to make equals the budget that the project manager may spend. Prioritization discussions are much easier to win. A shortage of resources for the project, for example, can lead to a material breach of contract situation, which most companies want to avoid. The environment in which customer projects are performed is much more public than in internal projects, and as much as

Table 2-1 Internal Projects vs. Customer Projects

	COMMON DIFFERENCES	
	Internal projects	Customer projects
Are . . . for the performing organization	Cost centers	Profit centers
Project selection is mostly made as . . .	Internal decision	Bid/no-bid decision
Project work for the requester is based on . . .	Internal agreements	Legally binding contracts
Team's familiarity with the target environment at project start is . . .	High	Low
Project managers are mostly . . .	Rather weak	Rather powerful
Obtaining resources is mostly . . .	Rather difficult	Rather easy
Management attention for the project is mostly . . .	Rather low	Rather high
Project managers must consider . . .	The interests of their own organization	The interests of both the customer and the contractor
Staffing and procurement are mostly managed by . . .	Functional units	Project management team
Budget is managed by . . .	Project sponsor	Project manager
Reputation inside their own organization is mostly . . .	Rather low	Rather high

contractor companies love to send information on successful projects to the media as case studies of their proficiency and expertise, they must also fear that the press will be informed of failures by frustrated customers.

Table 2-1 shows common differences between the types of projects.

In addition to these two common types of projects, there are two more that I have observed in my practice. These are much rarer, but especially in the case of capital projects, their value can be very high.

2.2.3 Capital Projects

This type of project is often found in infrastructure projects, which are funded with private money. A project manager in a capital project does not work for a sponsor but for investors, and the project is not performed inside an organization—the organization has instead been set up just for the project. An example is a Build, Operate, Transfer (BOT) project, in which the investor organization builds new or upgrades existing infrastructure, operates it, and creates income from this operation until a contractually agreed date is achieved (commonly after 30 to 35 years). Then, the ownership of the infrastructure is handed over to the state. There are some examples of successful BOT projects (e.g., the expansion of the Motorway A8 between Münich and Augsburg in Germany), but also of failures that are due to flawed business cases and damaging political influence (e.g., the Taiwan High Speed Railway Project[*]).

---

[*] (PMI, 2005).

2.2.4 "Razor-and-Blade" Projects (or Freebie Projects)

It may sound strange, but some companies do customer projects for free. For example, large logistics companies run integration projects to tightly link the internal logistics of their customers, mostly manufacturing and trading companies, with their own transportation and warehouse logistic systems. The benefit for the customer is the creation of a highly integrated and coordinated system that combines the requirements of speed, reliability, and efficiency as well as flexibility, depending on the system developed and the companies involved[*]. To gain these benefits, the customer pays a price: Once such a system has been put into place, and the operations systems of the two companies are interwoven and optimized, it is hard for the customer to switch to another logistics provider. The business case for the logistics company is analogous to the traditional business case of razor blade manufacturers: They give away the long-living razors at a low price or for free once and enjoy a continuous income stream from the disposable blades. The business model has also been applied to many other products, such as instant cameras, ink-jet printers, and mobile phone services. While they often struggle to lock in the customer, they are a relatively easy undertaking for the logistics providers, since they do not invest into the customer from a distance as a razor manufacturer would do, but rather integrate deeply into the customers' operations and day-by-day decision processes; therefore, they have good knowledge of the internal affairs of the customer.

Freebie projects seem rather exotic. They come with risks, such as failure of the new business of the customer to create the predicted demand for logistics services as expected and, as a consequence, insufficient business to pay back the investment. The business model seems unlikely to become a mainstream project type for the future, but who knows?[†]

2.2.5 The Same Methods for Different Types of Projects?

In most project management literature, no distinction is made between these fundamental types of projects and methods. Behaviors and approaches are commonly recommended for all types of projects—even when they have proven successful only for one of them and may be detrimental for others.

2.2.6 Conclusion

Most seasoned project managers are rightly proud of their experience, but many of them have only managed one type of project. Internal project managers often lack the experience of running a project as a profitable business against a legally enforceable contract. This contrasts with project managers from customer projects, who have not managed projects for stakeholders,

---

[*] The World Bank considers international logistics a core driver of economic wealth poverty reduction and develops a bi-yearly "Logistics Performance Index" (Arvis et al., 2014).

[†] A paper published by The Law School of the University of Chicago shows how this business model was not actively developed and implemented but emerged in a competitive market over a long time (Picker, 2010).

who will possibly remain their colleagues for years after the end of the project, and with whom an "after me the deluge" mentality ignoring the long-term future would not be appropriate.

Certain practices fit different kinds of projects, but others are more specific in their application. Chapter 3 goes into more detail on this question.

2.3 What Is the Matrix?

This topic is tightly linked with the previous one: Most projects today are not performed in isolation in a desert oasis or on a remote island but are done inside existing organizations—corporations, government agencies, associations, or other forms of incorporations. Most of these organizations are not optimized for managing projects but for ongoing operations, such as manufacturing, services, trade, and other forms of repetitive activities that can be standardized, cost-optimized, and sold to customers with inexpensive pricing. These organizations tend to "operationalize" every activity they do, which allows them to get cheaper, faster, and better over time. Operational repetition comes with many benefits, including predictability and order. It also allows for optimization when organizations and their personnel go through learning curves and helps them improve cost efficiency and quality. Operations can be very effective in using scaling effects, reducing costs of productions and services by increasing throughput numbers. For example, if it costs a manufacturer $1,000 to produce a certain widget, and each widget will cost in addition $1 for raw materials, making 1,000 widgets will cost the company $2,000 or $2 per item. Making 10,000 widgets instead will cost $11,000 or $1.10 per item; a cost advantage from scaling of 45%.

Many famous thinkers, consultants, and practitioners in engineering and business sciences have focused on repetitive processes in the past. Frederic Winslow Taylor showed in the early 20th century how work can be made cheaper and more predictive when it is broken down into smaller portions, each of them repetitive again but easier to understand and to optimize[*]. Influenced by his work, the Reichsausschuß für Arbeitszeitermittlung (REFA) was founded in Germany in 1924[†], an association that sent trained experts as *"Zeitnehmer"* (time assessors) into production facilities and later also into administrative environments where they used stopwatches and workflow report sheets to identify and eliminate time waste. It has been reported that in 1943, when German production in the Second World War was at a peak level, their number was at 12,000[‡]. An essential element of the Toyota Production System (TPS) developed in the decades following the Second World War is the principle that a process must be optimized before it gets automated; otherwise, in conjunction with efficient ones, inefficient processes will be also automated and will then be much harder to identify and improve[§]. Software consultants know the principle that business process engineering must precede business software implementation to make the best use of the software—or, in other words, that it is easier to adjust an organization to the specifics of a software product implemented than

---

[*] (Taylor, 1911).
[†] (Schulte-Zurhausen, 2014, p. 11).
[‡] (Hachtmann, 2008, p. 21).
[§] (Toyota, 2003).

vice-versa. This may be the most basic principle of operations management: Organizational structures mirror those of the business processes.

Frenchman Henri Fayol is another example how business thinkers of the early 20th century understood organizations as mainly functional, dedicated to operations with their ongoing and repetitive tasks, and his work is highly influential still today. In his "Administrative Theory", he focused on the influence of administration on the separation of work and the predictability of its results. Workers should have a clear and unambiguous line of command, should know their place inside the organization and should be allowed to rely on the stability of their position and work environment[*]. Some of Fayol's principles are applicable to project management, but the majority are not valid for our field, where most work is inevitably dominated by frequent organizational change, temporary assignments, and overlapping chains of commands.

Operations have another opportunity for optimization, and it is commonly overlooked that projects basically do not have this: the ability to utilize resources close to 100%. This opportunity is generally impacted by the non-repetitiveness of the workflow logic. For example: A worker, Jill, has been solicited from an outside vendor, and the vendor will bill 10 days for her. She works on an item that she must first process along three activities A, C, and E, with three days' duration each. Between these activities, another person, Jack, must do some work on these items along activities B and D. Jack may for example be her internal liaison, and it is his job to review whether Jill is on the right track with her work by auditing her processes and inspecting the results. It is not quite clear how long it will take Jack to do his reviews but the plan gives him two days for each of the reviews, so if each of Jill's activities will take two days as well, a simple workflow is created and shown in the swim-lane diagram[†] in Figure 2-4.

In the example, Jill has been booked for the entire duration; sending her away during her idle times between her activities bears the risk that she will not be back in time when Jack has completed his activities. Leaving her assigned to the project bears the risk that she will be billed for ten days but will only work for six. In operations environments, where situations are repeatable and predictable long-term in advance, solutions can be found and applied over time

---

[*] (Wood, 2002).

[†] For this type of diagram, rows (swim lanes) are drawn along a time axis for workers and activities are being located on them. It makes it easy to identify times during which a person is available or not. It further helps identify times when the resource is idle or is allocated to more than one task at a time. Instead of individuals, swim lanes can also represent equipment, meeting rooms, teams, contractors, and so on. Larger swim-lane diagrams generally don't show dependencies. They would become too confusing.

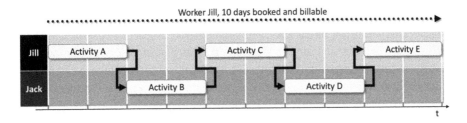

Figure 2-4 A flow of activities over two workers, Jill and Jack, who have to work alternately on the same item in two-day cycles.

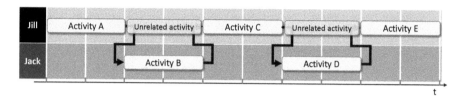

Figure 2-5   In operations, repetitive idle times can be identified and often filled with other work to keep resource Jill busy and productive.

to keep resources busy with other tasks. In the example, Jill could do some unrelated work and be kept productive outside the work flow described in Figure 2-4.

Figure 2-5 shows how operations can smooth resource usage to reduce billable or chargeable resource idle times—"billable" for external resources, "chargeable" for internal. Operations can plan work and book resources to keep idle times at an unavoidable minimum. Billable or chargeable resource idle times are considered waste, and operation managers are in comparatively a good position to eliminate waste over repetition.

The point at which no resource is idle is sometimes referred to as the "Pareto optimum". At this point, whenever one wishes to dedicate some resources to a new task, one has to take them away from another one, because the system has no free resources that can be used as a bench reserve for unforeseen needs for resources. Projects are rarely in a Pareto optimum. One of the reasons is uniqueness; the typical lack of repetitiveness in projects. Uniqueness may not be a characteristic of every single project task; some tasks may indeed be highly repetitive, but uniqueness is the characteristic of the project in its entirety. A second reason is the need for reserves to cope with the uncertainties that come with this uniqueness. If a resource, a person, or a piece of equipment or another kind of temporary resource is expected to work four weeks on a task to finish it, the resource will be most likely booked for a longer period, in case it should take the person or equipment longer to finish the task. Often enough, project work is similar to a journey that was planned by a person who intended to take the sprinter train which is travelling the direct way from A to B, but finds himself or herself later in a slow train, which is travelling along the winding and scenic river valley, slow enough to allow people make pictures of landmarks on the way, but not what was originally intended.

Reserves are generally like fire extinguishers: They must be bought and need to be inspected and maintained in fixed intervals, all costly activities for items that spend most of the time idle mounted at walls. In case of fire, people are happy to have them, but burning houses are not that frequent. Reserve booking times may also sometimes not be needed, but the booking periods are often fixed, and if in such cases resource run out of work, charging or billing will nevertheless continue. By the way: Burning houses may not be that frequent, but work finished late definitively is. Managing booking reserves is a core competency of a project manager, necessary to handle the uncertain elements of a project, and so is defending them against organizational requests to free them for other tasks. Having booking reserves will lead to idle times of resources and increase project costs; not having them will make the project flammable in case of difficult situations. A project manager without reserves is a feeble observer of great events.

In classes, I often have discussions with cost engineers* about whether billable and charge-able idle times exist at all. They fundamentally contradict costing principles they consider sacred, and one of them is that resources should be paid for work, not for availability. To take a simple example, think of rental cars: If someone rents a car for five days, but uses it only on two days, and the person approaches the rental company and wants to pay only for two days, the most likely answer of the company will be something like: "You had the car booked and available for five days, and this is what you must pay for". The company could not rent out the car to someone else during the five days and could not earn money from it during that time and bills the customer for the full five days. The company may be accommodating and may not charge the customer for the three idle days, but there is no obligation for them to do so.

In projects, idle times often come as a surprise. Rarely, another piece of work is due by such a time that the free person is qualified to do. So, smoothing the lag in the resource usage may not be possible, and sending the resource to another assignment may be risky. When will the resource come back? Another problem is that trying to fill billable or chargeable idle times in projects with unrelated work often leads to a disrupted work flow, which leads to delays and missed deadlines. Adhering to the correct order of work is an essential prerequisite for not fail-ing, and ignoring it leads to a multitude of problems, and possibly crisis.

The way for operations managers to achieve the goal of optimization is by separating them-selves from the uncertainties of the surrounding environment to the maximum possible. At the beginning of the industrial revolution, they erected big walls that surrounded factory build-ings to isolate them from the uncertainties of weather as well as social impacts. Later, those walls were supported by the steam engine that isolated productions from the capriciousness of wind, water currents, and exhaustible human or animal muscle strength that had commonly impacted operations at earlier times. An interesting anecdote in this context: When Lewis Wickes Hine, an American sociologist and photographer, published his first photographs of child labor in US factories around the year 1910, the public was shocked at the practices hid-den from it, and he was repeatedly threatened for making these practices visible[†]. The isolation also had a social aspect; the public outside the factory walls should not know and discuss the industrial practices inside them.

Still today, we see many attempts to isolate operational processes from the environment in which they are performed. An example are call centers whose secondary job is to satisfy customers when they have questions and complaints, but whose primary function is to keep operations undisrupted by questions, concerns, and complaints from customers. The isolation of operations allows for optimization of processes and helps improve predictability and order, but it separates the internal processes from their markets. In essence, isolation was and still is a means to make operations a closed-skill discipline (discussed in Chapter 1).

Isolation also has an inner-organizational aspect: Functional units—for example divisions, departments, or branch offices—tend to erect walls separating them from other units inside the organization and from their managers, as the perceived interest of the unit is often more important than the interests and the strategy of the organization as a whole. These walls are often intended to separate a business unit from the uncertainties in other business units to a maximum degree and to prevent a "weak spot" in one part of the organization from impacting

---

* They are called project controllers in some environments, while their job is much less to control the project but to monitor project cost development.

† (National Archives, n.d.).

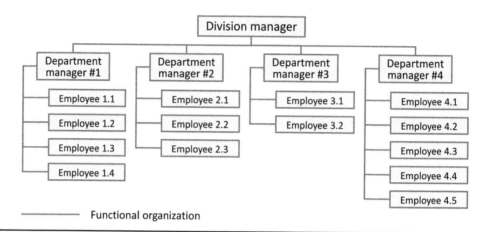

Figure 2-6 A typical partial line organization, which consists of a division that is broken down into departments with different head-counts (simplified structure).

the other parts. A major leader of the quality management discipline, W. Edwards Deming, demanded in his famous 14 points that organizations should "Break down barriers between departments"[*].

Now, let us see how such a functional organization may prepare itself to perform a project with its own resources. Figure 2-6 shows a division as a sub-organization inside a functional organization, which may be a firm, a government agency, an association, or something similar. The division is broken down into departments of different sizes.

There is an interesting aspect about the functional organization: In the example, it is very likely that Department Manager #4 is the most influential one among the department managers since this person has the largest department. Depending on the corporate culture and the people involved, there may be other influencing factors, such as who earns the money for the organization, whose role is the most business-critical one, who has most management attention, the largest office chair, or the fastest company car. There may be many other factors that positively influence the power of a unit manager, but head-count is generally a strong one. With more span of control, functional managers are able to obtain more budget responsibility. They have the potential to be more productive and business critical, and as stakeholders in business decisions, their business interests are regarded as stronger, and their arguments are more convincing. One should also not underestimate off-the-record influences: They have more ears and eyes to know what is chattered on the "grapevine radio"—the informal information channels in an organization—and they also have more voices to spread rumors. Gossip is generally most successful when many people repeat and confirm the same statement, even when this is false—an effect for which Norman Mailer once coined the term, "factoid"[†]. In factoids, evidence is replaced by consensus, and a larger department can create the impression of wide consensus much easier than a smaller one.

When a project manager applies for a new job, the first questions by the recruiter relates to project experience and formal qualifications. The first question to a functional manager relates

---

[*] (Deming, 1986, p. 24).
[†] (Mailer, 1973).

to the person's span of control and the budget managed in the current or previous job, and the numbers answered are regarded as an essential element of the person's overall qualifications. Functional managers commonly dislike having to give away resources, because their subordinates give them self-esteem, reputation, and power.

Figure 2-7 shows how such a functional organization would staff a project with its own human resources.

This overlay of lines of command is commonly referred to as a matrix organization. A matrix organization may be planned and "engineered"; more often, it just emerges. Matrix organizations in essence are not limited to the project/operations overlay; many other management domains are effectively cross-functional as well, such as cost engineering, quality management, data and privacy protection, work-place safety, and more. As these are generally operational by nature, i.e., repetitive and ongoing, the organization has time to arrange the multitude of power balances between the different management domains. The overlaps with the functional organization seem similar to that of project management at first glance, but these cross-functional management disciplines and project management could not be more different. Projects are temporary by nature, which means that time may be too tight to avoid, mediate, or reconcile potential conflicts. How project managers manage these conflicts effectively and engage functional stakeholders in their projects is often the decisive factor between project success and failure.

There is a corollary of the link between unit head count and perception of power and influence that can often be observed, especially in older and more rigid organizations: When the department manager in Figure 2-6/Figure 2-7 is the most powerful with the largest headcount, it is easy to understand what the person's first feeling will be, when he or she learns that a subordinate will be taken from the department temporarily to support a new project as a team member. The first emotion will likely be a perceived loss of power and importance. There are ways to overcome this perception, such as internal charge systems that give department managers budgetary compensation for the sacrificed workers. Increased influence of the business unit on the execution of the project and its outcomes is also a well-working justification for the temporary transfer of unit resources to the project. Participation in the success of the

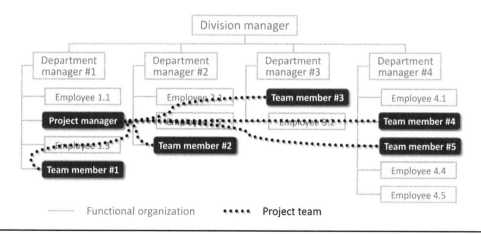

Figure 2-7   A matrix built from a division inside a classical line organization with a project performed cross-functionally across the departments; each box represents an employee of the organization.

project may be a third kind of compensation, but these benefits are secondary perceptions; the primary one is the loss of power and influence from the temporarily absent resource.

So, for the sake of the example, we have Jane, employee 1.2 in Department #1, who is also assigned to the project as the project manager. She has a team member, #2, Jill, who is also employee 2.1 supervised by Jack, Department Manager #2, and has just been recommended to go to Hawaii for five days and perform an important project task. Jane speaks with Jill and explains to her how important the task is for project success, asking her to book a flight, a hotel, and a rental car for five days. Hawaii is a place where Jill has never been before, and she would love to go there, even if it is just for work. She informs Jack, her department manager, who is not positively impressed; instead, he becomes angry. In the end, she is his employee and cannot be assigned for a task without his approval. He needs her at her work place and explains that there is a continuous demand by customers to talk with her. Given the time she will spend travelling there and back, and the time zone difference, she will not be able to meet her duties to the department during her absence, says Jack. His department provides an income stream for the organization, and he must therefore reject Jill's assignment. So Jill has to make a tough decision between her functional job and her project team assignment. Will she buy a ticket, book a hotel room and a rental car? Or will she follow the functional manager's strict "No"?

Jill is in a classical "Servant of Two Masters" dilemma[*], and as she cannot make herself available to both the functional manager and the project manager at the same time; she has to make a decision. In many organizations, it seems more likely that she will stay. Jack, the functional manager, has "disciplinary power" over her. He decides on promotion and incentives, conducts her performance reviews, and is in a position to fire her. Time is also an issue. When the project assignment is over, Jill will have to return to Jack's department and will have to work with him, her functional manager, again. It is also interesting to look into the past. Jill has known the functional manager for quite some time and knows what to expect from him, while Jane, the project manager, is a fairly new contact for her, and Jill has not developed any rapport with Jane so far. The result will be that Jane will not be able to send Jill to Hawaii, and the same will happen with everyone else she approaches. The project will be delayed, and it will be difficult to deliver all functions expected from it.

The situation described in this little case-story has a name in project management: we call it a "weak matrix"—weak of course as seen from the perspective of the project manager, whose power and influence are limited. Figure 2-8 explains the position of the weak matrix in a continuum of organizational matrix layouts between functional only and project only.

Is it then also possible that a project is performed in a strong matrix environment? Most customer projects are run in strong matrix settings. As it was discussed before, project managers in customer projects provide the income for the contractor organization, and if—in the example— no one flies to Hawaii, the contractor may not be able to send the next invoice to the customer. In addition, the contractor will possibly run into a breach-of-contract situation. Most organizations do not want to be in breach of contract, and part of the project manager responsibility is to

---

[*] Project managers should know the very funny Renaissance comedy, *"Il servitore di due padroni"* (Goldoni, 1753). Servant Truffaldino, working for a lady, is always hungry and decides to hire with a second master, a man, to earn himself a double income. Hired by two masters, he creates a lot of chaos and confusion, but finally he helps them to confess their love in public and get married. Servants of two masters can create confusion but can also connect people.

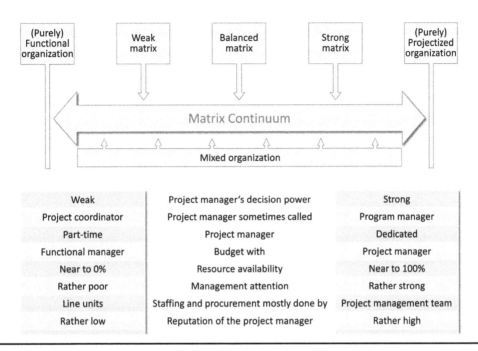

Weak	Project manager's decision power	Strong
Project coordinator	Project manager sometimes called	Program manager
Part-time	Project manager	Dedicated
Functional manager	Budget with	Project manager
Near to 0%	Resource availability	Near to 100%
Rather poor	Management attention	Rather strong
Line units	Staffing and procurement mostly done by	Project management team
Rather low	Reputation of the project manager	Rather high

Figure 2-8 There is a continuum between a purely functional organization and a purely projectized organization with different forms of matrix organizations in between. There are also mixed organizations, where all kinds of matrices can occur.

ensure that it will not happen. There are other situations in which one can find strong matrices. High-ranked project sponsors can strengthen project managers—unless they lose interest in the project, because they have found a new pet project—and projects that are performed to meet legal requirements. An example was April 15, 2004, when large publicly traded United States corporations had to become Sarbanes-Oxley Act (SOX) compliant, a set of rulings as to how corporations had to certify the completeness, correctness, and timeliness of information communicated to shareholders. (Smaller firms and foreign companies listed at United States stock exchanges had been given time until the end of the first business year that ended after June 15, 2005.) A staged set of provisions was implemented with the Act, which culminated in a ten-year prison provision for CEOs and CFOs for knowingly or willfully providing false certifications[*]. Most managers do not want to go to jail, so pressure was high on the organizations to finish their SOX projects on time and meet all legal requirements. Project managers in these mostly internal projects were generally powerful, because they had maximum management attention that could not deteriorate and had to work against hard deadlines imposed by law.

Between strong and weak matrices are balanced matrices, something that one can observe rarely. Even organizations that claimed to have implemented a balanced matrix, at a closer glance, had a dominance of either projects or operations, in most cases the latter. Balanced matrices do of course exist, and they seem the best solution for companies that run a major portfolio of customer projects. In practice, they are hard to sustain because even the most willing organizations will feel the disintegrating forces of power-games and politics from time to time.

[*] (AICPA, 2008, p. 13).

There are two more environments for projects: purely functional and projectized. The purely functional organization is made of departments that run projects without project managers, just with their own management resources, often in the form of "over-the-fence" projects, where departments work on a project, each of them with its own phase, and when the phase is considered finished, the project is thrown "over the fence" to the next department, which will be responsible for the next phase. I saw the approach some years ago used by a manufacturer of custom printing machines, a company that went into insolvency in 2011; and I am sure that the "over-the-fence" approach was a significant contributor to its bankruptcy, possibly the root cause. While it would have been desperately necessary for the company to have project managers that could overlook a project from the beginning to the end, the department managers did not accept the perception of demotion that came with such a role, and instead allowed the projects to be fragmented on a massive scale. The company made great machines and was profitable over decades, but when the market changed after the year 2000 and the demand for printed newspapers, magazines, etc. shrank due to the success of online magazines, the company was no longer able to adapt. One can hide organizational deficiencies under high profits, but when the profits evaporate, the faults become visible.

On the other end of the continuum is the purely projectized organization. These organizations are found in large-scale construction and infrastructure projects, but also in R&D, when an organization is just founded for the purpose of a specific research project. The organization may be a consortium, a joint venture of contractors on time in which each venturer provides resources for the project, or a special purpose vehicle (SPV), which concentrates more on the aspects of project management and project finances, and procures project work and deliverables from outside the organization. The discussions in this book assume that most projects are run inside matrix organizations, the extremes described here are highly interesting but will not be discussed in further detail.

Figure 2-8 above also includes mixed type of matrix organization described, called the "composite organization". This is a type of organization where several projects may be run concurrently in different matrix situations. An example may be a development division in a company, which is performing projects for its own purposes and others for external customers. In such settings, it is often observed that project managers in customer projects have a stronger standing since they create income from which a budget can be calculated, and they have to ensure compliance with contractual obligations. Another aspect of a composite organization may be that the prioritization and management attention of projects changes over time. Another type of manager is what I like to call an HBR manager: This person reads the *Harvard Business Review, Fast Company,* or another great management periodical. Every other month, they surprise the organization with a new idea and a new project that needs to be performed; and all ideas are good ideas, but the organization is not capable of coping with both the speed and the scale of change, and worst of all, these managers cannot dedicate enough attention and support to all the projects at the same time. HBR managers can be found in many organizations today, and with all their good intentions and ideas, they can cause true havoc. When the projects that have been started with high expectations and the promise of strong management attention, and active support to the project manager lose the attention of senior managers, those projects may experience unexpected delays and cost overruns. Then the corporate project traffic lights are all set to "red", and the project gets back the attention and prioritization it lost, possibly in conjunction with a new project manager.

2.4 The Economics of Attention

Attention is rarely described as a resource in literature, but looking at common definitions of the term resource, attention definitively is one. It may be the key resource of all. Management attention is not necessarily a guarantee for the availability of other resources, including people, equipment, material, funding, etc., but without management attention these other resources will be difficult or even impossible to obtain. The attention of other stakeholders may be important too. It can lead to support and positive engagement but also to resistance and conflicts. Managing attention is based on the way the project manager and the team communicate with stakeholders, and this is another example of "navigating between two monsters". On one side is the insufficient communication, when stakeholders are left unclear on the project, on the costs, and other kinds of burdens that it will impose on the organization, its progress, its risks and problems, and the other things that the stakeholders should know. One may instead communicate too much, and stakeholders then may feel that their time is being wasted in meetings and with long reports that they are expected to study and comment on. Over- and under-communications are threats that can both lead to a loss of attention by management and other stakeholders.

Another risk is poor communication, ignoring the diverse interaction and behavioral needs of people from different cultures, which may lead to misunderstandings and wrong expectations. One more menace for projects may be the chemistry between the people involved; since communications between people with poor rapport is rarely effective, the contents of the communications may be corrected but the emotions that overlay them can not be. The chosen style of communications can also become a problem. James Pennebaker[*] described three different styles called formal, analytical, and narrative. The formal style may feel dry, stiff, and demanding and relates to status and position as much as to the contents of the communications. Used as evidence at court, formal communication may be strong, as this documentation tends to be briefer and "crisper" than other styles. The analytical style communicates honesty and the preparation and ability to deal with complexity. Frequent distinctions are another common element of analytic communications that their backers often show by pigeonholing and classifying people and things. Texts are often longer to account for explanations and clarifications. For users of the narrative style, the favorite method is story-telling. They show a lot of social competency and awake empathy in others as much as they have it by themselves. Some people are good in all three styles, others strongly prefer one or two styles, which tend to be those they master most successfully.

A project progress report, written for managers, a customer, or for an internal requester in a narrative style may be entertaining and successful, or appear to the readers as a waste of their time. A request for the allocation of resources written in highly formal style may be successful when it makes the impression that standing processes are followed, or fail when a functional manager feels the basic lack of empathy that is typical for the style and responds that apparently his or her needs are being ignored. With a strong analytic style, a project manager can help stakeholders better understand problems and their underlying causes or appear as a know-it-all to these others. When project managers wish to ensure the attention of managers, heavy-handed best-practice approaches will not work; project managers need

---

[*] (Pennebaker, 2011, pp. 79-82).

susceptibility for the dynamics of success and failure and for the sensitivities and preferences of the stakeholders involved.

As a trainer, I get asked frequently what I consider a good frequency for meetings and how long they should take. I cannot honestly answer this question as it depends too much on the specific situation. There are project situations where it is beneficial to have a short meeting a day, maybe in the form of a stand-up meeting in the morning, as a basic briefing for coordination and discussion of issues and concerns. In other situations, it may be necessary to have sufficient session time to discuss various aspects of a matter and resolve disputes without rushing, something one would not want to do too often to leave team members and other stakeholders sufficient time to do their work.

2.5 How Project Managers Learn

A student of mine recently told me how he learned of his assignment to his current project. A manager of the organization met him unintentionally—or so it seemed—in the company restaurant and told him, "Good that I am meeting you here. Have you heard that you have been selected for the new project xyz? A great project with massive impact. The future of the entire company depends on it—it will have top priority. To not create false expectations on your side, there will also be a shortage of resources in the coming months, as we have other top-priority projects concurrently, some are approaching their hot phases, and budget and funding may also be insufficient, but we both know that you are able to successfully run such a key project even under these difficult circumstances and deliver it in time against a . . . well . . . challenging deadline. The project is highly important, and to some degree, the future of the entire company depends on it; have I said that already? I will arrange that you get the details of the project ASAP so that you can start immediately creating the first results. By the way, the project is extremely important, as I said, but please, do not let your daily work suffer from it. We do not want competitors to laugh up their sleeves when we are not prepared to tackle them on the market place. I am aware that at this moment, I should ask you whether you are prepared to take over the project, but I think we can agree to skip this formalism. The project must be done, and you are the best person to whom I can entrust it. So, let me explain the details . . .". It is probably needless to say that the manager did not have any details, just some basic information he could communicate.

When taking over the assignment for a project, the project manager's knowledge about it is typically near to zero. The project may be an internal project, and managers have made a decision to add the new project to the portfolio, actually a classical investment decision. In some organizations, there is a formal process with a project request and approval procedure, which includes an impact analysis and an assessment of resources needed versus those available. The majority of internal project decisions are probably rather made spontaneously and ad hoc, and cases such as the previous example, where the project manager is informed in a more private setting, are not rare at all. In 2012, I was asked by the PMO of an engineering company to help them set up a project selection and prioritization process for their internal project portfolio. We had the full backing of the organization's CEO for the development of a process intended to make sure that the company did not have too many projects at a time, that the projects were better coordinated, that load balancing assigned projects to units that had availability and not to those that were already overstrained, and to also make sure that the future project managers would benefit from a documented project evaluation process. This documented process would provide them with records that would give them a better understanding of the original goals and requirements of the project and serve as a basis for initial decision making. Then the CEO found out that as a consequence, he would have to adhere to the formal process in the future. The development was then stopped, and project managers in the company still had to take up assignments with almost no documentation from the selection process as a first source of information.

Documentation in internal projects from project selection may be near to zero; in customer projects instead, its quantity may be overwhelming. If the project is a customer project, there may have been a process beginning with a request for proposal (RFP) or an invitation for bid (IFB) delivered by the customer at the beginning. In most cases, a response would have been sent to the customer in the form of a bid, proposal, or another kind of quotation, and after some clarifications, negotiations, and discussions, the customer would award the contract to the contractor. Then a project manager was assigned, who often has not been part of the business development process and now has to scan through all the paperwork to find the information that helps to quickly make the first decisions.

In both cases, it is common to call the project manager into the project only after the decision has been made. The recommendation is strong to include project managers in the internal decision process whether to start a project, or in the business development process in customer projects to validate decisions made for soundness and realism, but this is rarely done in practice. Project managers are more often expected to focus on the delivery side and many are not free and available to also support the project selection processes with practitioners' experience and expertise. Even if they were part of the internal project selection process or supported the development of the business with the customer, it is likely that the knowledge that they could gain is limited. They were not in the formal position of the project manager and had access only to information that was easily accessible, possibly to no information at all. Even asking people from the customer organization or internally for details on the current situation, something that is relevant to understanding what is needed for the change or the development that the project needs to bring about for the future, they may not have been able to obtain in-depth information, because they are not yet project managers at all. These and other causes make taking over a project a true adventure with a large amount of uncertainty about the issues that lie ahead. If there is a moment at all that the project manager will have full knowledge, it will be at the end of the project. The unique nature of the project makes it a major learning experience, and every new project will add new knowledge in technical dimensions, as well as in social, interpersonal, organizational, and even political dimensions.

The ability and preparedness for lifelong learning is therefore a trait that effective project managers have in common—they never consider their understanding complete. This ability goes together with situational intelligence, which consists of three elements: (1) the understanding that the same practice that was successful in a given situation in the past may fail in a different situation, or vice versa; (2) the ability to adjust practices to the specific needs of the project and the current situation; and (3) the care that this adaptiveness is not perceived by others as signals of lack of authenticity or reliability. See Figure 2-9 on this learning process.

The following are other developments that occur naturally while the project proceeds along its lifecycle:

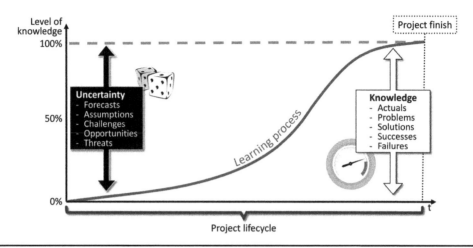

Figure 2-9   Each project goes along with a learning process for the project manager and all other involved stakeholders, which leads the team from a status of uncertainty to one of knowledge (repetition from the first chapter).

- The number of feasible options for decision making decreases. Decisions and work made in the past limit decision options for the future. Late in the project, there may not be enough remaining budget, time until the next deadline, and resources available short-term to implement certain decisions.

- The costs of implementing decisions increase. With sufficient time, you can choose a cost-effective solution. Under time pressure, the only options available may be expensive ones. Simply think of a package that you need to send to somewhere. Consider the cost when the there is no rush in the delivery and the cost when the package needs to be there the next morning.

The time to assess the impacts of decisions on the project and to accurately document the decisions to better manage these impacts gets shorter and shorter. The risk of unintended consequences of decisions is highest when they must be made as "field decisions" during implementations.

The project manager is not alone with this learning process, which is shown in Figure 2-10. Virtually all stakeholders go through such a learning and development process, during which they realize what the project is about, what its results will look like, how the team members and contractors cooperate, and many other factors. Uniqueness comes with novelty, and novelty comes with learning. It is interesting to take a closer look at the dynamics of this learning process.

The American psychologists Chris Argyris and Donald Schön discussed two types of learning[*], which they called "single-loop learning" and "double-loop learning". Their model refers to organizational learning and interpersonal behavior, and I wondered if it can give insights into project management learning processes and behaviors. I think it does.

They look at three elements of decision making and actions in situations that are influenced by uncertainty and therefore by learning processes. They researched how people, groups, and

---

[*] (Argyris & Schön, 1978).

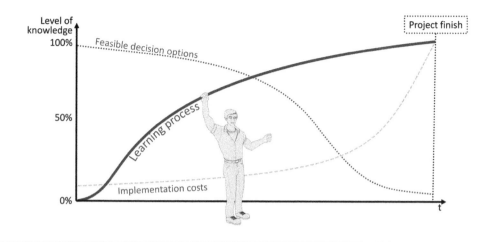

Figure 2-10   Proactive project management methods try to enhance the learning process, so that knowledge is available early, when many options for decisions are still available, and when they are less costly to implement.

organizations reflect their decisions and actions and assessed which were good and which were, judged with the advantage of hindsight, rather poor. They then assessed what measures they should take to respond to the deviations between the intended outcomes and the actual ones. In their theories of action, they described three elements of this process: governing variables, action strategies, and consequences. Translated into project management, one can describe these three elements of a learning and decision-making process as follows:

1. **Management conjectures.** The governing variables—factors that influence the way project managers, their supervisors, and their teams want the project be performed as a whole, as well as the individual activities, that combine to become a major endeavor. The most obvious of these factors are plans and also the strategies that underlie these plans. Very religious or ideological persons may have dogmas that drive their behavior, while for scientists, paradigms can serve a similar purpose. Management conjectures are based on the past or on the current, but they relate to the future, which has the uncomfortable characteristic that it has an element of uncertainty. So, these conjectures are a combination of guesses, assumptions, forecasts, and objectives, as well as of expectations and agreements. Even contracts are conjectures as they guide actions based on the expectation that others—monitored or not—will adhere with their obligations, and on the fact that most people and organizations do not want to run into a breach of contract situation. There is no complete contract, a contract can therefore not remove all uncertainties, instead contracts must leave many details open for later clarification, just like plans do[*].

2. **Implementations.** The conjectures are then implemented in the form of managed project activities, which are kinds of action strategies in Argyris and Schön's words. The implementation may refer to actions of the project manager, the project management

---

[*] (Hart & Moore, 1998).

team, or the project team as a whole. This step relates to the present day, and while some plans are more accurately implemented than other plans, the results will mostly contradict the conjectures to a smaller or major degree. The future cannot be predicted with 100% accuracy.

3. **Outcomes.** In project management, it is unlikely that outcomes will be precisely as expected. Deviations from the conjectures will occur relating to the deliverables created, delivery dates, costs, and stakeholders will learn from these deviations.

To give a simple example: A team member Jack, a developer, has estimated that he will realistically be able to finish some development work after three weeks for a project A. Based on the estimate and a five days schedule contingency, Jack's activity has been scheduled for four weeks' duration, and this duration has been included in the project schedule. Based on the schedule, Jack got booked, and the costs for the booking were added to the cost plan. Information on Jack was added to the project's human resource plan, and his communication needs and those of Jill, his manager, were documented in a communications plan. Before Jack can start the work for project A, he will be assigned to another task in a project B, and delays there may influence his availability for project A. Delays may also come from dependencies of his task with other tasks that must have been finished and provide input to his work before he can start. While the schedule has been based on the assumption that these delays will not occur, the uncertainty that the assumption may be wrong has been added to the risk register. Once Jack has finished the activity, he will work over a month for project C and then come back and do other work on project A. Everything has been well planned, management conjectures have been made clear and documented. Two weeks after Jack starts working on the activity, he identifies some unexpected difficulties in the task that will lead to a required working time of eight weeks instead of the originally estimated three weeks. At the end of the four weeks booking time, he will be assigned to another project for four weeks, and only after that will he be able to return to the activity and finish it. As Jack cannot be easily replaced by another developer once the work has been progressed so far, and as no other developer would be available that quickly as a substitute for him, the activity finish will be delayed by two months and Jack will be blocked from doing other work that he was scheduled to do after his return from project C. One should also note that the additional work on the activity creates additional costs for the project that lead to a certain budget overrun.

There are two kinds of lessons for the project manager from the situation:

- **Single-loop learning.** The project manager puts the blame on Jack. Jack had promised to finish the activity after three weeks, and the project manager had even given him an extra week as a schedule contingency, and this is still not enough. The project manager requires Jack to take appropriate measures to finish the activity in a timely manner before he leaves the team to help project C and to protect the budget from the overrun.

- **Double-loop learning.** The project manager seeks the error in his or her own plan. Maybe the uncertainty of the activity duration was underestimated. Maybe the activity was not described in sufficient detail, or not enough effort was spent to ensure the quality of the results of the predecessor activities, so that Jack could do his work based on them without difficulties that take him time to resolve. This approach may also include looking at Jack's future assignments, and those of other team members as well, and adjusting the plan to avoid repeating errors such as this one, and it will lead to discussions with some key stakeholders if the budget is still sufficient to finish the project successfully.

Outcomes in projects are generally not 100% as expected, and while a deviation may be a problem for the project, it is also a new lesson for the project manager and other stakeholders that reality provides to them. There are projects that follow mostly the planned stream of actions and results. There are also rollercoaster projects with breath-taking ups and downs, making plans obsolete in just moments. Both kinds of projects entail requirements for project managers and other stakeholders for learning, which is challenging, but also an opportunity to grow. Figure 2-11 shows the two learning loops for project management that are applied in the example.

Management conjectures include everything that influences the way project managers and involved stakeholders want the project performed. They may include not only values and mental representations of the world around, but also systems of rules and recipes that these people have adopted. Some people even ask astrologers and augurs for advice on the decisions and actions. Guiding presumptions relate to the future, but they are based on extrapolating things that are considered true for the present or the past. As mentioned earlier, highly religious or ideological people often have dogmas to which they adhere closely. In sciences, paradigms can serve a similar purpose. In businesses "best practices" and strategies often serve this purpose. In project management, all these may play a role, but the foremost guiding set of guiding presumptions is found in plans. Project managers model their projects in the form of scope statements and work breakdown structures, time by using schedules, costs in the form of budgets and cost baselines, and so on—all kinds of plans they make to give the future of the project a degree of predictability and order and to be able to communicate this future to the stakeholders involved. Once the project has been initiated—i.e., it has been founded and the project manager authorized—these plans are the starting point of further development.

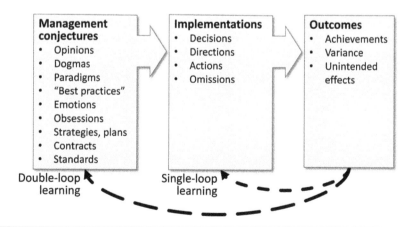

Figure 2-11   Learning from the outcomes of an organization's or a person's actions can take two forms: single-loop learning, which addresses the implementation, or double-loop learning, which addresses the governing conjectures of management.

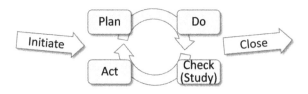

Figure 2-12   The PDCA cycle, adapted for project management.

Figure 2-12 illustrates how the well-known Plan-Do-Check-Act (PDCA) cycle is generally implemented in project management. It was originally developed by Walter Shewhart and popularized by Deming[*] for usage in operations, especially in manufacturing environments. For project management, two more elements are generally added to PDCA: initiating and closing. These elements adapt the model to the fundamentally temporary nature of project[†].

Combining Argyris and Schön's model with the PDCA cycle gives two options how project managers can learn from the differences between their plans and what is happening in project reality, as shown Figure 2-13.

Another example for the model shown in Figure 2-13 is a statement that a manager of project managers made to me some years ago when we discussed a change request management section that should be part of a project management seminar, and that he wanted to see removed: "I cannot expect from my project managers that they modify their plans again and again". His company ran projects implementing metal machining equipment, and I asked him what his project managers should do if a project was late and on the way to miss a deadline. He was very firm that the project managers should accelerate the project by adding further resources ("Crashing"[‡]), by de-scoping, or by using other means. All these approaches have limitations: more resources may not necessarily accelerate the project, but will definitively

---

[*] Deming in his books referred to the cycle as Plan-Do-Study-Act.

[†] Compare this model with the model of the *PMBOK® Guide* (PMI, 2013), whose central process groups are essentially the same: planning, executing, and monitoring and controlling.

[‡] The term is probably derived from the "Crash courses" that some British language schools offer for managers, who only have a few weeks to learn a language and need to make fast progress in a highly focused one-to-one setting.

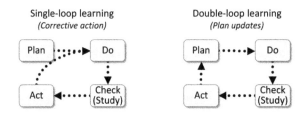

Figure 2-13   In project management, an implementation of Argyris and Schön's theory of learning is the distinction between corrective action and updating the plans. Using the Deming cycle of "Plan-Do-Check-Act" helps visualize how the learning types are applied in project management.

increase costs, and de-scoping is constrained by the contract. In addition, replacing an expensive "Item A" with a cheaper "Item B" without documenting it, possibly even without informing the customer, leads to scope creep and to discrepancies between the documentation, which describes the original intentions, and the actual implementation. In the example, single-loop learning describes attempts to accelerate the project and bring it back to schedule. Double-loop learning—in project management—means rethinking and reworking the plan. It may include talking with the requester, customer, or any other stakeholder who defined the deadline and discuss re-scheduling. Depending on the specific project situation, each approach may be appropriate—or wrong.

The example confirms another statement of Chris Argyris, who once noted that double-loop learning is far more difficult than single-loop learning*. A project manager applying single-loop learning puts the blame on someone else: workers, contractors, bad luck, etc. The plan was good, but people, business partners, and the reality of an unsupportive environment made it impossible to implement it. Double-loop learning in project management means seeking the error in oneself. The project manager is responsible for the plan, and if the plan doesn't work, the project manager has to assume this responsibility. Generally blaming others is arrogant. Generally seeking the fault in oneself is depressive. Project managers should avoid arrogance as well as depression and seek the truth somewhere in between, basing their learning style—single- or double-loop—on the available information and on the possibility that they have made a mistake or someone else has. The basic lesson from Argyris and Schön's model is how learning is linked with mistakes, and how a learning culture essentially stems from a culture of accepting mistakes as human.

So, each project is a sequence of new lessons to be learned, which contribute to a learning process that spans over the entire project, and during the different projects that project managers perform in their life, they collect further knowledge. Lifelong learning—probably more from the mistakes that we have made, and also those that we have observed by others, than from successes—is an essential element of a project manager's professional development process. The dynamics of success and failure are dictating that this process will never end as long as we are members of the discipline.

Project managers have a third option, in addition to corrective action and plan updates: Disregard the plan. In my classes, I occasionally hear statements like: "In my current project, I found out that the plan did not work anyway, so I decided to ignore it and told my team members to simply do their job. In the end, it does not matter if a cat is black or white, as long as it catches mice. Once the result is there, no one will ask how it was attained". One could call this zero-loop learning, and in most cases, it is bad practice. If the project does not need a plan, the project manager should not have wasted resources developing one right from the start. If the project needs a plan, because it may not be enough that everyone simply does his or her job, the team may do the wrong things and the project may fail. This statement describes the worst learning cycle. One may of course correctly raise the objection that the learning process results in the finding that a plan, which the project manager originally considered necessary, is later found dispensable because the project is much simpler than originally thought, but this is not the situation communicated in the statement above. It is rather a message from a person, who could not cope with the management tasks assigned, and who

---

* (Argyris, 2006).

rejected the learning process, which comes with every new project, as too strenuous. Project managers that prefer this style often assign themselves attributes like "goal-oriented" or "result-driven". I recommend to be careful with them. Rats also chase mice, and if the mouse in the house is caught by a rat, this may not be good news at all, because the small problem has been replaced by a larger one. Project managers who are not prepared to go through any learning cycle may cause more problems than they resolve; they do not grow in their professionalism from project to project, and on the way to obtaining their results, they may cause havoc to the performing organization and the customer.

2.6 Game Theory for Project Managers—A Brief Introduction

A case-story: Around the year 1990, I managed a project, which was part of a major program performed by an automotive company. A new production line was to be implemented, and 124 projects needed to be finished on time to meet a start of production (SOP) deadline of the principal program and its deliverable, the production line. The SOP was the most critical moment of the entire program, and all project activities in the program were managed to make this event happen on time. The program included projects for the construction of factory buildings, delivery and installation of infrastructure and machines, development and implementation of electronic hardware and software systems to control the production, recruitment and organization of workers and administrative staff, and development and implementation of environmental protection arrangements. One project was mandated for the integration of all these components to become functioning as a highly complex system. Most projects were run by contractors, only a small percentage were performed as internal projects by the automotive company. The good news was that out of these 124 projects, 121 were able to meet their delivery dates, which computes to a success rate of 97.6%, something one may consider quite impressive. Three projects were late: two performed by contractors and one internal project. Actually, those three projects were truly a problem, and they jeopardized the start of production. In a situation as described here, there may be a small number of rather uncritical projects, where the program does not get delayed too much by a delayed contributing project; but the vast majority of projects are time-critical, potential show-stoppers, for the program, and so were the three late projects here. They were critical enough to delay the SOP, which meant a fiasco for the program. The output from the production line was already envisaged in the company's production plan, staff had been hired and trained to work on the new line, and shareholders expected that the increasing production would also increase the value of their stakes. The three late projects were not a 97.6% success rate, they meant a 100% failure to meet the deadline and the business case of the program.

It would have been beneficial for the program manager to have the full knowledge of the expected problems early to be able to make decisions at a time, when there were still many feasible options left and when these options were cost-effective to implement[*]. The program manager may have been able to inform senior managers early and find solutions to limit the damage to the program. But will the project managers of the late projects tell him that their

---

[*] The correlation between early knowledge and decision making was discussed in Chapter 1.

projects will deliver late? In reality, this is more unlikely than likely. I had later opportunities to speak with them and they told me in detail how they perceived the situation.

First, it took all three project managers quite long to understand how late their projects actually were. They did not use methods such as network diagramming that would have enabled them to develop useful forecasts as a basis for scheduling and early booking and would have allowed them to validate the contractual delivery dates. Most of the time on their projects, all they could do was guess when certain events would take place, and at what time what work will be done, based on experience (they were very proud of their experience). For progress assessments and decision making, they had to rely on statements from team members and subcontractors. This may have been sufficient in other projects, but not for this program, which was dependent on reliable deliveries of built-to-order products and timely performance of related services. Additional pressure came from managers in the contractor organizations, the internal sponsors of the two customer projects. They asked so often, "Will you deliver on time?" and the project managers answered so often, "Yes, of course!", that they finally believed it themselves. Without a plan to measure project progress against, the project managers asked their team members and subcontractors, "Will we deliver on time?" and the answer they got was generally, "Of course we will. Why not?" Many organizations use traffic light symbols to communicate project status: Green if the project is on plan, red if it is not. In the example, the projects became classical melon projects: green on the outside, but the deeper one drills, the redder they get. As soon as the project managers became aware that they did not have options at hand to deliver on time, they should have informed their managers, and it would have been the contractual and moral obligation of the contractor organizations for whom these project managers worked to inform the customer.

There was also an internal project that was late, which was performed by another department, and basically the same happened with the internal project. This is not just a contractual issue, but a common problem of project managers who focus on technical matters and ignore organizational concerns.

When they finally became aware of the lateness of their projects, they found solace in the insight that it was unlikely that they were the only ones with late projects. Part of their experience was that a complex program like this one had never adhered to schedule, and at least one late project has always led to a missed program deadline. All they needed to do was wait for someone else to raise the finger and tell the program manager that their project was late. Then, this other project manager would have all the problems. The person's employer would possibly have to pay damages or penalties to the customer and suffer from the loss of reputation that such a case brings, and the customer would tell all other contractors that they no longer have to deliver on time. This is how to let a problem resolve itself by just waiting. One cannot be a good project manager without experience, but experience is a teacher who sometimes teaches the wrong lessons.

At one point in time, not long before the SOP date, one of the project managers indeed had to give in as he could no longer hide the truth that his project would not be able to meet the deadline. Facts became obvious and the program manager realized that the program results would not be delivered on time. The SOP got delayed by three months. Next, the program manager informed the project managers of all other projects involved that the program had been rescheduled and that their deliveries were now expected 13 weeks later. For the other two of the three late projects, this message was a blessing. All their concerns and worries vanished in a moment, and they had no difficulties meeting the new date. They were the secret, but nevertheless happy, winners in the game.

Experts in game theory call such a situation a "chicken race". The term goes back to the 1955 movie *Rebel Without a Cause,* in which two young men, Jim (James Dean) and Buzz (Corey Allen), drive stolen cars over the edge of a sea-side cliff with a simple rule: They have to jump out of the moving car shortly before it races over the cliff, and the first who jumps out of his car is the chicken, the coward.

Races like that actually exist in real life. When the young lads do not have a cliff available, they may instead race on a road toward each other, and the first driver who veers off is the chicken. If no roads and no cars are available, the daredevils may wait on railroad tracks—preferably high-speed rail tracks in Europe or Asia, where trains can run 300 km/h (187 mph) or more—and jump off the tracks in the very last moment. Again, the one who jumps first is the chicken. Some jump late.

The term "game theory" may sound somewhat cozy, playful, and amusing; however, it is a serious mathematical discipline that has strong implications for other sciences, including biology and sociology, and also in political, economic, and business sciences. The situations that it helps us understand are generally not cozy at all. Game theory is often described in highly cryptic notations, but the situations that it depicts are pretty normal experiences that people make in everyday life. They are often driven by limitation of resources and time and by people who can choose between cooperative and noncooperative—competitive[*]—behavior in situations, when cooperation would serve the interests of the community of players, but the competitiveness would serve the individual interest. Limitation of resources and of time are two typical aspects of project managers, so project managers should understand how it is applicable for their discipline as well. I will try to avoid cryptic notations and build on the common sense element that underlies game theory so deeply, and will also provide practical advice from it to help you avoid or deal with some of the conflicts that almost inevitably come with managing projects.

A core term of game theory is the "Nash Equilibrium". It describes a malevolent kind of "invisible hand", a stable balance of forces that does not automatically lead to an optimum for a group of players, but instead makes it impossible to achieve this optimum because players follow their[†] perceived particular interests that conflict with group interests[†].

An example of a Nash Equilibrium is the massive over-fishing of the seas that is currently done on a global scale. Major fishing companies today rapidly destroy not only the property of all mankind and the economic basis for coastal communities, but also the resources for their own businesses, to their own long-term detriment[‡]. The owners of the fleets of fishing boats are well aware that their over-exploitation of fish stocks is not sustainable, but they are nevertheless

---

[*] It sometimes leads to confusion that the use of the word "competitive" in game theory does not necessarily relate to a setting that includes running a competition but is used to describe any kind of noncooperative behavior. In this context, "competitive" stands for any kind of egoistic behavior that is focused on the particular interests and on meeting private agendas, in contrast to cooperative behavior which tries to meet common interests and agendas of the "players".

[†] The concept of a benevolent "invisible hand" was invented by Adam Smith in his book *The Wealth of Nations,* arguing that import tariffs are unnecessary, as people generally prefer to buy local goods over buying from foreign sources to limit business risk (Smith, 1776). Meanwhile, reality has repeatedly shown that this statement is not always correct.

[‡] (United Nations, 2004).

in a race to be faster in this exploitation than others. The awareness of the limitations of sea life, the basic resource their operations require to survive, does not make them more careful about its conservation, but more exploitative and competitive in applying highly effective technologies, and while they are increasingly investing in larger and more powerful vessels and equipment such as sonar and trawls (fishing nets), they accelerate the process that will finally lead to the break-down of their own businesses. Their dilemma is that if any of these organizations would change its mind and turn to more collaborative and sustainable practices, others may not do that, and this organization would fear becoming the loser in the business competition. This is the essence of a Nash Equilibrium: "It will not do any good if only I change my behavior as long as others are not changing their behaviors with me, so why should I do that?" Scientists and politicians tried repeatedly to manage the situation by limiting the amount of fish the fleets may take home and ban the most destructive practices like bottom trawling, but mostly to no avail. The fishing quota decided by politicians from time to time are much higher than what scientists consider sustainable*, and with the lack of accurate real-time monitoring and enforcing mechanisms, one may doubt that fleet owners will adhere with them anyway—the financial incentive of not following rules is another Nash Equilibrium.

A further example of a Nash Equilibrium is the arms race during cold-war time in the 20th century. It was expensive and risky for both parties, USA and USSR, and while most people considered the nuclear balance a guarantor of peace, reality was that in the year 1983 alone, the world evaded an almost certain nuclear catastrophe twice, more by luck and insubordination and heroism of individuals, than by the mechanics of deterrence that many people at that time considered infallible[†]. The fear of losing the arms race was stronger on both sides than the understanding that common interest of both nations would have dictated to negotiate an end to it, focus public money on investments that are beneficial for both societies, and save both countries (and the rest of the world) from the abyss of mutual nuclear annihilation. An equilibrium of commonality would have been to meet and find ways to avoid the arms race, but the Nash Equilibrium, dictated by particular interests over joint interests and also by serious paranoia on both sides, was stronger, until the 1983 incidents showed that both countries were following the wrong way. After these incidents, common sense politics returned, and agreements on disarmament were made during a slow process, which was mostly driven by the need to build mutual trust strengthened by mutual monitoring systems. We have since perceived how hard it is to avoid the return of mutual distrust and with it the Nash Equilibrium, which can lead to a new cold war at any time.

The core topic of game theory is the conflict between the particular interests of persons, organizations, possibly even countries involved in a "game", and the common interests of these game players. There are also applications for animals and even plants, as they share commonality in the dynamics that lead players into dilemma situations. Game theory gives strong clues on how to identify these dynamics and respond to them early. Sometimes, it helps find resolutions when these dilemmas occur, but often, it does not. In the discussion of the learning process during a project earlier in the section and how this process correlates with the loss

---

* *The Guardian* described an example in December 2014 (Harvey & Neslen, 2014). It is interesting how politicians proudly emphasized that they were only interested in the short-term profit of the national industry, not in the interests of fisheries as a whole, and did also not consider the question of sustainability of fishery business.

† A brief discussion of these incidents can be found at Wired.com (Beckhusen, 2013).

of feasible options and the increasing costs of the remaining options, the loss was described as steady over time. In game-theoretical dilemmas, all options often get lost just in a moment. We may be able to avoid dilemma situations in projects and elsewhere if we identify their emergence early; if we do not, we may no longer be able to rectify the problems that may arise. Game-theoretical dilemmas are often gridlocks that make it difficult or even impossible to get out, once they have developed.

The "chicken run" in the movie *Rebel Without a Cause* is a fictitious story, of course, but it is based on real events, and it also gives an example for this rapid loss of options: Buzz, one of the two young men driving the cars over the cliff, finds just before the edge that a strap on the left sleeve of his leather jacket is stuck in the door handle. He cannot open the car door to jump out. There is not enough time left for him to free the strap or stop the car, and he crashes with the car into the ground at the bottom of the cliff.

If in 1983, either nation during the Cold War had fired the first missile, no country would have had an option left except to go for mutual annihilation. When a final rush for the last fish will be over because all fish have been taken from the oceans, the last option for the big fisheries will be formal insolvency. When people in projects or programs run into Nash Equilibria, they may no longer have the options left to make decisions that could lead them out of the dilemma and bring the project back on the tracks that lead to success. In the example, the project manager worst off will be the one who did his or her job and communicated the delay first, not those who did not communicate it at all.

The original concepts of game theory were developed in the late 1940s by Oskar Morgenstern and John von Neumann and further developed in the 1950s by John Nash[*]. Many others have contributed to the field since then, and the theoretical concepts have been tested repeatedly in labs and observed in reality by scientists. Experts in game theory were hired for high-value auctions to increase the chance of success for bidders—generally without much success. As one should note, game theory is neither alchemy nor an ideology, but experts in game theory are nevertheless often well paid as consultants in highly competitive situations. Game theory can help make dilemmas avoidable, but it does not provide a magic potion when the dilemmas and gridlocks have evolved or will inevitably occur, as in auctions[†].

The focus of this discussion is not game theory as a theoretical and scientific discipline, but its situational application in project management, so the following case stories may help you understand the dilemma situations we often face as project managers. Most are straightforward and understandable; game theory is not rocket science. It is therefore astonishing how often project managers and other stakeholders repeat the same mistakes and run into the same troubles again and again, and how little attention is given to steering clear of Nash Equilibria, when basic project decisions are made.

2.6.1 Two-Players' Games

The most popular example of game theory is the so-called "prisoner's dilemma". Two prisoners, John and Jane, are believed to have jointly committed a crime; both have been arrested and are

---

[*] A movie was made on John Nash's life in 2001 under the name *A Beautiful Mind* with Russell Crowe starring as John Nash.

[†] Especially in public auctions, cooperative behavior may be considered illegal bid-rigging.

Table 2-2 The Expectations of Jane and John for the Time They Will Spend in Jail When They Confess

	Jane will serve	John will serve	Total time in jail
No one confesses	½ y.	½ y.	1 y.
Jane confesses	0 y.	3 y.	3 y.
John confesses	3 y.	0 y.	3 y.
Both confess	2 y.	2 y.	4 y.

questioned by the police about their participation in the crime in separate rooms without the ability to communicate with each other. Each has to expect two years in jail. The investigators offer each of them immediate release on parole as a reward if the person confesses, but only if both do not confess. If one person confesses, the punishment would be increased by one year for the other person. Without any confessions, the investigating prosecutors will not have enough evidence to build a lawsuit and will have to let the prisoners go. They have six months to produce the evidence, during which they can keep the prisoners in jail on remand.

An overview of the numbers in the example is given in the table in Table 2-2.

The cooperative solution—among the prisoners—would be that both remain silent and not confess. This would lead to a total time in jail of one year, the sum of the six months that each of them would be jailed on remand before they would have to be released. The likeliness is indeed higher that both will confess the crime, lured by the hope for immediate release and also to avoid the three-year verdict when the other person confesses. This will then lead to a two-year sentence for each of them and a total time in jail for both of them adding up to four years.

The official notation in game theory is shown in Table 2-3.

As far as I have observed, project managers do not find themselves that often in prisons; the lawsuits that they are involved with are more likely to deal with contractual matters. Nevertheless, they find themselves in comparable dilemma situations. You may be more familiar with the following case-story (based on a true story that I was in direct contact with in 2010).

Frank has been assigned as the project manager to a major internal software project, whose entire technical part was to be performed on fixed price by a contractor organization called CatDog*. This was a consortium, a temporary joint venture, of two software companies, Cat, Inc. and Dog Corp., who were teaming up for the project. Each of the companies had a 50%

---

* This is, of course, not the real name.

Table 2-3 Jane's and John's Dilemma in Game-Theoretical Notation

		John	
		cooperates	competes
Jane	cooperates	½ – ½	3 – 0
	competes	0 – 3	2 – 2

share in the CatDog consortium, which was founded solely for this project, and was planned to be closed once the project was finished. The teaming decision was not based on a business decision elaborated by the two venturers, but to meet a firm requirement of the customer organization, who wanted to combine the different competency areas of both companies. The project was undertaken to develop and implement a custom workflow control system for a manufacturing environment, a field, in which Cat, Inc. already had a lot of experience and a well-established reputation.

Dog Corp. had a long history of projects for the specific customer but was new in this technical field and hoped that the project might help the company develop know-how for future projects. Dog Corp. further hoped that the successful project would open doors for the company into future business by providing a working solution with a satisfied customer that could be utilized in future as a sales reference. During the project, a difficult decision needed to be made by the consortium: whether to add some new and supplemental functionality, which would increase the performance of the solution and its ability to meet customer requirements by a quantum leap. Both companies understood that this change would add significant costs to the project and following the fixed-price contract, the consortium would not be able to bill these additional costs to the customer. The change would have added further risks to the project and probably consume the project's profitability.

While the decision was more and more delayed by the two venturing companies, Frank observed that tensions were mounting between them, and that these tensions already led to a general decrease in the performance of the consortium and to a degradation of their intermediate results. He had read somewhere that "Organizations which design systems are constrained to produce designs, which are copies of the communication structures of these organizations" (Conway's law[*]), which can be simplified to "Systems reflect the relationships among those who make them", and wondered now, if the performance and quality problems were just mirroring the eroding relation between the two companies. He spoke with team members from both companies in private and got a confirmation of the deteriorating relationships that he observed. During his personal talks, he could also identify the basic cause for the breakdown of communications (and of the team spirit) inside the contractor consortium: the completely different business interests of the two companies. Cat, Inc. was well established in the field of the project, and while the happy customer had some value to the company, the profit from the project was considered more important. Dog Corp., in contrast, was interested in a powerful and convincing showcase and strategic reference from the customer, which would have helped get access to technologies and a new market; a market, by the way, in which it may have to compete in the future with Cat, Inc., who did not feel comfortable about this outlook. A cooperative approach by both companies would have been to make a joint decision, which would have required both parties to give in with the other company, at least partially, and meet halfway, something that neither company was prepared to do. The result, a change decision that took too long was detrimental to both companies and also to the project: The poor performance and quality had already led to rework, which affected the profits and led to a dissatisfied customer. The dilemma situation can be seen in Table 2-4.

It was hard for Frank to bring the parties together to discuss the issues without exposing his confidential sources that gave him the deeper understanding of the situation, and a lot of distrust and hostility had meanwhile built up that were also not easy to fix. His advantage

---

[*] (Conway, 1968).

Table 2-4 The Dilemma of the Two Companies in the Project

	Cat, Inc., gets	Dog Corp. gets	Total benefit
Both give in	Profit	Reference customer	Profit, reference customer
Cat, Inc. gives in	--	Reference customer	Reference customer
Dog Corp. gives in	Profit	--	Profit
No one gives in	--	--	--

was that the conflict was still fresh enough to not have run into a full and unresolvable dead-lock, and that the parties also felt uncomfortable in the situation and were happy that a third party offered help. When he had the full understanding of the problem, he convinced his management to agree to some financial incentives to the consortium, as well as promise it a professionally written endorsement letter. Another promise was to contribute to a case study article written for a special interest magazine. For the customer, this was a less expensive solution than suffering from delays and having to correct poor quality later, and the customer was actually interested in the value-adding change. The situation was finally resolved in the interest of all parties.

Two-players' games are similar to chess, boxing or football, where the "players" are actually the teams. Their basic problem is the direct confrontation between two parties, each of them trying not to lose, and winning often includes weakening the opponent. The Cold War arms race mentioned above was such a two-players' game. It was not only a nuclear arms race, but also an attempt to weaken the economy of the other party to make it finally give up. The confrontation can happen because of conflicting interests, but also clashing egos or fear may be a problem. Often, one finds a mix of these factors. In retrospective, the dynamics under way are generally easy to understand, but forecasting the emergence and escalation of these dilemma situations can be difficult. To make things worse in project management, the attention of the project manager, the project management team, and of governing bodies is mostly consumed by a multitude of commercial, technical, interpersonal, and organizational issues, and the emergence of dilemmas is often noticed late. When one finally notices the problems that have occurred, it may be too late to resolve them.

2.6.2 Multi-Players' Games 1: The Tragedy of the Commons

Some years ago, I was asked by a manufacturer of custom machines, a company I am calling here Lion Ltd., to help analyze why a portfolio that consisted at any time of 40 to 50 concurrent customer projects had much less commercial utilization of their engineering staff and equipment as the company considered feasible. Lion's focus was indeed more on the utilization of the people rather than the equipment, as they were the most dominant cost and time factor in their projects that consisted of the development, construction, dispatch, and commissioning of their machines. With the diversity of skills that these people used for the projects, the flexibility to allocate an idle team member to another task was limited; a developer of control software, for instance, could not be directed to weld a steel construction, and vice versa. The company considered it possible with good planning to ensure that their staff would work in productive and billable assignments for over 85% of their working time, but the actual rates were rather around 60% to 75% and at times, even less. The company had a long backlog of work to be finished, and the high proportion of unproductive and non-billable worker time—mostly for rework, but also for idle times—put the workforce under a lot of stress to do the same work in less time. Recruiting and training new staff would have helped reduce the stress, but would have been an expensive and timeconsuming activity, and with the long backlog, the company lacked the free capacities, as their workforce was busy to work for the next deliveries. As the utilization of current personnel was lower than desired, such a decision would have been difficult to justify, and management had indeed imposed a hiring freeze, which further increased the pressure on people. The performance problems also diminished the profitability from the projects. Damage claims, penalties, and missed contractual incentives caused by frequent, late deliveries reduced profits, and quality problems that needed to be fixed promptly when they occurred caused additional costs. To make things even more difficult, the higher the pressure from the work load rose, the more the utilization of their staff fell. Lion's management believed that the company could be performing much better, but did not know how, except by applying additional pressure on staff, which made things worse.

I conducted a series of interviews and soon, the picture became quite clear: There were several causes that together led to the poor utilization, but the most prominent was the company's internal charging system in which project managers had to "pay" fixed amounts of money for each hour that a person worked in the project; the money was then transferred to the business unit that provided the person—the organization ran in essence a classical profit center model, which distributes the profits over the projects and the units providing the resources. The total profit of the organization was the sum of the individual profits of its parts, and when all parts are making profits, the organization must do so as well, management assumed. If one of these parts, a project or a business unit, had a loss, this was easily identified, and the focus of executives would then be directed to finding and resolving the problem at this "weak spot". The project managers had to book the workers some weeks in advance, but the sums charged were not calculated based on the hours booked, but on actual working times. It was considered normal, in adapting to the uncertainties in the company's business, that booking times were longer than expected working times. When a worker was forecast to work eight weeks on a set of tasks, the person was commonly booked for nine weeks, giving the project a one-week booking reserve, which generally seemed reasonable to cover risks and which was considered in the calculations of the hourly rates of the internal charging system. The workers would use unused booking times for administrative work such as putting together their expense sheets or writing internal reports. Of course, Lion Ltd. had a sophisticated Enterprise Project Management (EPM) software solution to manage the allocation of resources to projects and the internal charges that came with them, and the system was customized to support the charge model across the corporation.

In times of pressure, the system got out of balance: Resources had to be booked earlier, at a time before start and end times of assignments could be reliably forecast, to avoid that necessary people and equipment would not be available at the times, when they were needed. The workers for the eight-week task needed to be booked for ten weeks, or even eleven, to cover the uncertainties that came with the early booking. As these workers were not shown as available

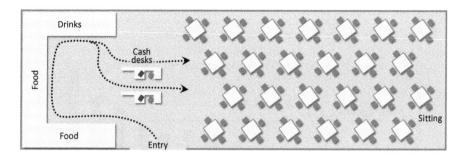

Figure 2-14   The arrangement in the self-service restaurant for customers in the furniture store. The dotted arrows describe the ways that customers are supposed to take from the entrance along the food and drink counters and the cash desks to the tables.

by the ERP when other project managers needed to book them, the other project managers also had to book earlier in advance and had to cover their additional risks with extended booking times to not be left without resources.

For an outsider, the dilemma situation that the project managers ran into may be difficult to understand, but another, much simpler case-story example from a completely different area may illustrate the basic dynamics that had occurred at Lion:

A large furniture store has a self-service restaurant for its customers with two areas: one to pick up food and drinks, and another one with tables and chairs for the customers to have their meals. There is a cash deck zone between the areas. Figure 2-14 shows the arrangement that most readers will find familiar. The dotted arrows describe the ways that customers normally use to enter the restaurant, take food and drinks, pay, and then find a seat.

The furniture store opens at 8:30 in the morning and a small number of customers go directly to the restaurant, as they come to the store only for an inexpensive breakfast. It takes the majority of customers at least 30 minutes to walk through the store and arrive at the restaurant. Most customers do not come alone; they come with their spouses, friends, and families and want to sit together with them at a table. The bottleneck resource in the example are the restaurant tables, and it is interesting to observe over some time how the tables are productively used, which means that there is at least one person at a table who is consuming food or drinks he or she has purchased.

Figure 2-15 shows that until 10:30 a.m. the situation is easy to describe: Someone who has paid at the cash desk and needs a place just occupies a free table and there are enough tables available. After 10:30, things change. More and more customers stand between the tables with a tray in the hands and do not find a place to sit even though not all of the tables are occupied by consuming customers. At some tables, there is just one person sitting without anything to eat or drink. These people sit there to reserve the table for the family that is on the way to get and pay for food and drinks. When the demand for free tables exceeds the supply, the number of people who reserve tables increases, and they will feel that they are doing that in support of their families. When it takes a family 60 minutes to finish a meal, the table is unusable for around 70 minutes.

The utilization for tables deteriorates further when waiting lines build up in the food zone and at the cash desks at around 10:30. Then, it will take the families longer to pass the process

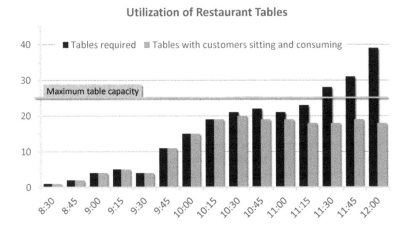

Figure 2-15 Number of tables required for eating and drinking, and the number of tables actually used for the purpose counted on a Saturday morning to lunch time.

and arrive at the table, and the waiting time for the reserving person will increase, thus further reducing the number of utilized tables. Then, a family may reserve and use a table for over 75 minutes, bringing the utilization further down to 70% and 75%. This poor utilization becomes quite a stable effect; it is another example of a Nash Equilibrium. One hundred percent would be a desired optimum, but the balance of forces veers the utilization down to the Nash Equilibrium, similar to a malevolent invisible hand. People waiting at otherwise empty tables reserved for their families to come with the food reduce the number of free tables at the time when they are needed most, and the shortage of tables makes it more imperative for these people to reserve tables. At the same time, other customers stand there with the trays in their hands and do not find an unoccupied table, while their meals get cold.

Back to the 50-project portfolio of machine manufacturer Lion Ltd., in which similar dynamics reduced the true utilization of corporate resources. Project managers could build availability reserves by booking needed resources for longer periods than actually necessary, or even book resources that they may not need at all, "just in case . . .". Resources were then either booked exclusively on one project, especially when work needed to be done on customer site, or were expected to work for multiple projects simultaneously, which led to massive over-booking, in one case of 800%. When it came to the moment that the unrealism of the resource booking became obvious, short-term decisions needed to be made by managers on the allocation of people to projects, and the resources went to the project that was in the most troubles. A problem for the company's staff members in the situation became the continuously high levels of distress in their work environments. This distress caused people to rush their work, neglecting the diligence that they would have normally applied. The managers believed that they also had cases of active disruptions and even sabotage, but they could not verify the allegation and could not relate it to specific people. This led to an effect known in psychology as effort-reward imbalance, which is a common cause for burnouts and in turn leads to absenteeism and staff attrition, an effect that Lion Ltd. could also observe and that increased the resource shortage further. Another effect was simply that people under time pressure are making more mistakes than people who have the time it takes for them to do their job. The entire organization was in an out-of-balance state that was not sustainable.

This kind of game-theoretical dilemma is often referred to as the "Tragedy of the Commons"*. Commons in this context stands for common land, mostly meadows and forests, which is not owned by a single person, but by the members of a community such as a village or a county. Commons were usual in historic times, but still exist today. An old name for such common land is "Allmend", which is still in use in parts of Germany, Switzerland, and Scandinavia and means "property of all". Common land is open for use by all members of the owning community, and in literature, one often finds a romantic transfiguration of the freedom that commons promise, but the reality is often rather grim. The Tragedy of the Commons refers to the observation that meadows, forests, lakes, etc. that were considered commons were generally in much worse condition than those that were in private hands. A pasture that is open for many people to graze their cattle needs some time to recover when the animals have finished grazing and have been taken away. On private ground, the owner will probably allow the grass time enough to grow to a size that makes it valuable for the cattle. On unregulated commons, when several people are entitled to let their cattle graze, it is rather unlikely that the grass will have enough time to grow, because each cattle owner must fear that, if he or she waits too long, another one will make use of the grass. That is why common pastures were mostly in much worse condition than private ones, and the same is true for community-owned forests, lakes, vineyards, etc. To avoid the Tragedy of the Commons, communities then tried to regulate their use, but found it difficult to come up with rules that were acceptable for all community members. Commons have always been a bone of contention, often vicious contention, as in the German peasants' wars in 1524 and 1525, when 75,000 peasants were killed who wanted to defend their commons and their freedom†.

Some of my readers will be aware of the *Leatherstocking Tales*, a series of novels by James Fenimore Cooper. The first novel, *The Pioneers,* deals with the regulation of a common good, a forest in the state of New York‡. In the novel, Cooper points to another aspect of the Tragedy of the Commons: The pioneers stage a pigeon hunt (Chapter XXII of the novel), during which they make a sport of shooting bullets into dense flocks of flying pigeons, not with the intention to hunt meals for their families but out of pure joy of killing. The pigeons—not owned by anyone and therefore also considered a common good—became victims of a meaningless mass killing in a world that was no longer theirs but was seized by the pioneers. A similar leitmotif can be found in Shakespeare's tragedy *King Lear,* where highly competitive actors are rather prepared to lose the common goods—the love of a father, an entire kingdom—than to cooperate for the sake of this good and risk losing as an individual. The last refuge that the king can take is madness. Human (and sometimes non-human) behavior in game-theoretical situations can be mind-boggling and damage people's basic ability to make common-sense decisions, consider matters of sustainability, and develop empathy for the needs and feelings of others.

In the example above, discussing the portfolio of customer projects at Lion Ltd., the commons were the corporate human resources that the project managers could use, or, to be more accurate, the times that they were available. Project managers could massively overbook them as reserves to respond to uncertainty. It is generally good practice to protect deadlines, funding

---

* (Hardin, 1968).
† (Virnich, 2013).
‡ (Cooper, 1823).

limitations, resource usage, and other constraints by placing reasonable reserves in the form of time, money, scope ("nice to haves"), etc. Another type of reserves are the booking reserves in the example, the extension of time slots that resources are made available on top of the expected time that it will probably take them to finish a task, or the booking of otherwise unassigned people as "benched players", to have them available if they are needed. Reserves of any kind come with costs to the organization, with the result that booking reserves excessively, sometimes referred to as "padding", can turn providing them into a bad practice.

I previously used the example of rental cars, that must be paid for based on booking times, not actual working times. If you book a rental car for five days, but use it only for two days, you still have to pay for the full five days. The same is mostly true for external staff: when a service provider has been contracted to provide human resources for four weeks, but those people have finished working after three weeks, the customer will likely have to pay the full four weeks—of course depending on the type and the clauses of the contract. Most resources are paid not for working time but for the time that they are available for the project, which means that they are not available for anyone else to use. The provider of the human resources or the car rental company may agree to a good-will agreement and not bill the resources for the whole time, but—depending on the contractual terms—it is rather unlikely that the customer can rely on this amiability.

At Lion Ltd., in contrast, the internal charging system allocated internal expenses to the project based on pure working time, so project managers could block resources far beyond the foreseeable needs without too much consideration of whether it is likely that they will need them. It is similar to the restaurant visitors mentioned previously. Lion Ltd. made the entire resource pool a common pasture for its projects, and when the exaggerated blocking of resources lead to reduced utilization, the company then had two options: Ignore the blocking of resources that normally comes with bookings and allow for over-allocations, which then (for some team members) added up to 800%, or try to find alternative work short-term for people who were booked but not needed, which was often impossible, given the specific skills that these people had, and also given that many were on remote customer sites and could not be easily taken to other locations. Such actions had another side effect; they gave team members the perception of being pushed around instead of making them feel committed to their project.

What was the solution? First, some short-term measures were taken to give back the perception to the employees that they are valued and their performance is esteemed. Some of these measures were quite simple and inexpensive, such as talking more with them and having a dinner together, but managers had to invest time and energy and needed to listen to their concerns and recommendations. Then, the internal charging system needed to be changed to be based on bookings instead of being based on actual work. This may seem simple, but it was not, mainly for two reasons:

1. The new practices for internal charging led to the need to make major changes in the software that the organization used. The default in the software was that chargeable resource costs were calculated as working hours X rates per hour, and the customization work needed was significant[*].

2. There was a lot of resistance from the project managers. The argumentation was similar to that of the fishing industry opposing quota discussed by scientists and politicians

---

[*] I noted earlier that it is easier to adapt an organization structure to a business software than vice-versa.

that would limit their "Freedom of the Seas", which entitles them to destroy the fish populations on which their business depends. Charging projects for booked resources motivated project managers to apply more active risk management to better understand uncertainties.

In addition, Lion Ltd.'s PMO installed the position of a resource administrator, who took care of effective allocation of people and major equipment to project tasks to avoid both over- and under-allocation, to the degree that this is possible. The combination of software, organizational decisions, and interpersonal action brought the desired effect. One should also say that some project managers needed to be replaced by new personnel, because some could not accept the changes to the practices, which required from them to not only consider their projects but also the effect on the organization when they allocated resources to their project tasks.

2.6.2 Multi-Players' Games 2: The Dilemma of the Concurrent Investments

I observed the next case story in a software project for the replacement of an existing solution with a new one. I will call the company who performed the project in the following discussion Tiger, Inc.; describing game-theoretical dilemma situations is rarely pleasant for the stakeholders involved and portrays them often in unflattering light, which is why using avatar names is preferable in this context. The company had contracted over 75% of the work to two contractors, Bulldog Ltd. and Shepherd, Inc., the latter being my customer. Bulldog and Shepherd then gave some work to subcontractors, and Shepherd's subcontractors in turn used third-tier subcontractors—one of them actually a second-tier subcontractor to Bulldog. The result was a complex multi-tier supply network as shown in Figure 2-16.

Multi-tier supply networks as this one pose some special challenges to project managers. They occur across all industries in growing numbers and complexity, and in practice, there are much more complex ones. The observation that their frequency and complexity is increasing is probably an effect of the still increasing speed of technological change in most industries. Corporations can often not develop the agility to cope with this change alone and, therefore, need the support of smaller organizations. A subcontractor may even be a single person as a champion to help the organization adapt more quickly to new opportunities. The most extreme example that I have been in contact with so far was the development of the Boeing 787

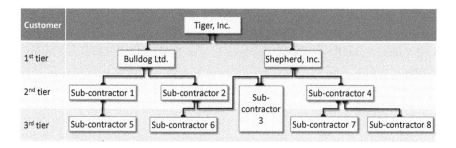

Figure 2-16   The supply network in the project of Tiger, Inc.

Dreamliner passenger aircraft. The major share of the development work was assigned to contractors and subcontractors over several tiers, among them many companies with whom Boeing had no contractual relationship: A legal doctrine called "Privity of contracts" in countries with Anglo-American Common law, "Protection of third parties' rights" in Civil law countries, says that only immediate contract partners have direct obligations against each other. A customer has no contractual relationship with a subcontractor.

To make things even more complicated, work was outsourced to companies in Europe and Japan, countries whose legal systems (mostly Civil law), business cultures, and work ethics were fundamentally different from what the development staff of Boeing was familiar with from previous business. The sales personnel of Boeing was much more experienced with these differences and could have given good advice to the developers on how to deal with them, but in a functional organization made of silos surrounded by almost impermeable walls, it is rather unlikely that communication works across the divisions: Sales people have to sell, developers have to develop, and none of them has the time and the will to talk intensively with the colleague from the other unit. Silo organizations are also a classical playground for Nash Equilibria, and the Dreamliner project was burdened with them internally and in the supply network; resulting in delays, massive budget overruns and technical problems, the worst of them being burning Lithium batteries, that even led to grounding of the aircraft in January 2013[*].

Tiger, Inc.'s project to develop and implement a business software was done by a multi-tier network of contractors that was difficult for them to manage. In the case of Subcontractor 5, actually a sub-subcontractor, Tiger's management was not even aware that the company worked for its project at all. They believed that Subcontractor 1 would do the work with its own resources, but their contract with Bulldog Ltd. did not exclude further subcontracting, and as the contractor lacked its own resources to do all the work, they secretly teamed with another company.

The supply network was widely managed by a system of contracts, which were mostly developed by legal staff, not by people from project management. Contracts rule the minimum obligations for each party involved that they have to meet to avoid stepping into a breach of contract situation. Contracts are rather a means of competitive than of cooperative action, because they are mostly useful when a conflict occurs, and when the parties understand that the conflict may finally end at court, another place of competitive rather than cooperative action. Contracts cannot obligate parties to show motivation and enthusiasm for the project, and they do not provide a foundation for intensive team motivation and mutual trust. I already cited Conway's Law, which says, "Organizations which design systems are constrained to produce designs which are copies of the communication structures of these organizations". In essence, the rule says that in order to produce performance systems that meet a multifaceted set of requirements of a customer, users, and other stakeholders, a project needs a performant team, which communicates in a style that is constant, consistent, effective, and intensive. Whether such a team actually exists in a project is mostly seen in the difficult moments, not in the easy ones.

In the case of the project of Tiger, Inc., the system of contracts worked fairly well and supported the interests of the project as long as things went generally as planned. Problems

---

[*] A battery fire occurred still in August 2015, two and a half years after the incidents that led to the groundings. This is an example of the difficulties of fixing problems that have their origins in the interfaces between actors in supply networks (Davies, 2015).

emerged, when unexpected situations arose during critical implementation activities that made it necessary to perform field changes and then to deviate from the plan. Field changes are *ad hoc* decisions, mostly during implementation phases, that need to be made in urgent situations. Field changes generally come with additional work and costs and may make extended operational disruptions necessary. They also come with risks, as they mostly need to circumvent change control processes—there is simply not enough time to assess the impacts of the change request in sufficient detail before making the decision. In Tiger, Inc.'s software project, the greatest problems were the interfaces between the various components, which in turn were the work items of the various contractors and subcontractors, and also implementing new software with the existing legacy systems inside the company, many of which had been used for years, some even for decades. The new software was developed in stages with growing complexity that was validated on a safe test system, where only some minor problems became visible that were easily fixed. The real problems occurred when the software was installed on productive systems. From the point of view of each of the contractors and subcontractors, they had met their contractual obligations and delivered a working component, and it was the job of the other contractor to fix their software. The fear was growing among the contractors and subcontractors that they had to put more work into their part of the software that the customer would not be prepared to pay for, that they would lose time and block their developers, who would then be not free to move to another customer project as planned. As each of these suppliers waited for another to invest additionally in the functioning of the full solution, the additional investment to make the full system work was not done at all. The last one to learn of the deficiencies of the software was the customer; every contractor reported that its work was going well, based mostly on information from the reports from the subcontractors, and the customer assumed that if every vendor did its job according to contract, they all should arrive at a working solution at the end.

To make things worse, a habit of mutual finger-pointing and recriminations evolved between the various companies. For each of them, it was clear that they had performed as expected on their contract and another company should be blamed for the interface problems. The dilemma started as a mathematical calculation: If I invest more than necessary, the additional costs will be on my side, but the benefit—an effective software product without quality issues—will be with the customer. Contractors and subcontractors soon developed a collective system of fundamental mutual distrust. The contracts did not really help; the customer had no direct access to the subcontractors because of the "Privity of Contracts" doctrine mentioned above*, and also because of the fear of liability claims, which made the contractors and subcontractors even less cooperative and communicative among each other and with the customer.

The "Tragedy of the Commons" describes a situation in which the Nash Equilibrium entails the concurrent exploitation of a resource without considering its attrition, possibly its destruction. The "Dilemma of the Concurrent Investments" describes a Nash Equilibrium that leads to important investments that will not be made. If each member of a group of players has to make an individual investment, but the benefit from the investment will be shared by all group members, a group member may follow a strategy to reduce its own investment and still enjoy

---

* In a multilevel supply chain or supply network, each party has only a legally enforceable relationship with its direct supplier or customer, but not with the supplier's supplier or the customer of the customer.

the share of the benefits from the investments of the other members. If all group members follow this strategy, there will not be any benefits left to share and enjoy. In the example, each
piece of work done by a player on top of what is contractually required is an individual investment in the common good, the software solution. Contracts in project management tend to get
outdated over time, which is not only due to changes in the project, but also to changes in the
environment in which the project is performed. If they cannot be updated in a timely manner,
they then need to be commenced, possibly even replaced with good will and the preparation
to consider the common good more important than the individual one. If this behavior is not
shared by all parties, the cooperative parties will feel that they are getting fooled by the non-
cooperative ones and may change their behavior too and become competitive as well. Then,
the Nash Equilibrium will occur because of the conflict between the particular interests—do
not invest more than what is inevitable—and the joint interest that each party gives its best to
contribute to a great project and build the solution that the customer needs.

2.6.3 Hope for Our Projects

Discussions on game theoretical dilemma situations may leave people unsettled. Similar to a
classical tragedy, they often occur as a one-way road leading into disaster without an alternative
escape option. Many respond to the dilemma situations by increasing pressure on their own
personnel and contractors, trying to enforce cooperation and communications, but in most
cases, this strengthens the deadlocks instead of resolving them. Authoritarian and competitive
behaviors are rarely solutions to dilemma situations. Another response may be to withdraw
from all conflicts and ignore the course into the dilemma affecting the project. Then, project
managers often focus on technical matters or respond to problems with hollow formalisms,
when they should actually become active to resolve organizational and interpersonal problems
and leave the solution of technical problems to their team members. Project managers who
tend to withdraw or avoid conflicts rarely have successful projects as they lack the will to pre-
emptively avoid the dilemmas and make needed changes when the emerge.

A simple observation may help find the solution: Humans are, to my knowledge, the only
life form in which individuals voluntarily join forces to carry heavy things together*. Before the
domestication of animals such as cattle, donkeys, horses, camels, and even elephants that could
carry goods or pull carriages or sledges, human muscle power was probably the only force that
could be used for the transportation of goods. Combined muscle power of hundreds or even
thousands of individuals built the Egyptian Pyramids, Stonehenge in England, the ancient
temples in Greece, and the temple of Angkor Wat. This force is as strong in making people join
and achieve common targets as the force of a game-theoretical dilemma that separates them
and makes them place individual interests over common ones.

Take, for instance, a group of three carrying a heavy washing machine upstairs to a fourth
floor apartment in an old city building without an elevator. The task is not simply a matter of
muscle force, but also of coordination: Staircases in these buildings are often narrow, and the
carriers have to arrange themselves around the load so that they can handle it jointly without
dropping it. The ends of the machine may differ in weight, and so does the power of the carriers, so they will arrange them in a way that power and weight will be best balanced. They

---

* Well, almost the only life form, some ant species do that as well (Gelblum et al., 2015).

have to coordinate the movement upstairs to avoid one person stumbling, so actually, they communicate a lot by feeling the movements of the load that they are carrying and adjusting the force that they apply to the load accordingly. They also have to communicate verbally to coordinate their movements, and even persons who are normally reticent become very talkative when they carry heavy loads with others. The minimum communications that one will hear is "Heave-ho!", which helps the carriers find a common work rhythm that makes moving the load easier. They have to observe and communicate exhaustion to decide upon breaks or to replace a fatigued person with someone who isn't worn out. The group has to be careful and not damage the washing machine or the walls of the staircase. Another part of the coordination is that the group does not want to carry the washing machine down to the basement or up to the fifth floor, when the job is taking it to the fourth. Coordination includes the common understanding of what is meant when the "job is done and is finished". It is also interesting how quickly (most) people can arrange themselves to build a working team for such a challenging job. People who have never met before just pick up the load and carry it up, mastering a task in a group that looks simple on the first glance, but requires a lot of coordination and communications.

Getting a job done as a contributing member of a functioning team is an experience that fills most people with joy and satisfaction. People are often more afraid of feeling neglected and lonely than of getting fooled by others. One can observe children learning to collaborate; two or more pick up a heavy branch fallen from a tree or a plank from a construction site and carry it to a place where they can play with it. No one told them the rules of collaboration; they just did it. You may need education and training to learn the modes of effective collaboration, but the basic ability and the joy of acting collaboratively seem to lie in our genes. Most people get a good feeling from being a member of an achieving group and share pride of the achievements with the other group members. People inherently want to belong to kinds of groups that exhibit the traits of a "Band of Brothers"*.

To understand what it takes to turn a project team into a Band of Brothers and Sisters, it is interesting to observe not-for-profit organizations and their projects, where people's work is mostly based on volunteerism and on the concept of acting for something that they consider a higher purpose. Volunteers in associations are comparable to people who work overtime without having a paid job. I have been a volunteer since 2002 for Project Management Institute (PMI), a global association with over 450,000 members†, and am currently the President of the PMI Münich Chapter, where we have counted, at times, up to 80 volunteers working for the operations and projects of the chapter, dedicating private time and energy to the association. In the chapter, they organize regular regional meetings with interesting speakers, run a monthly magazine, manage a bi-yearly congress, support projects for social purposes, and do many more activities. The management of the chapter is also done by volunteers, and the most activities of PMI as the principal organization, such as standardization and certification, are also performed by volunteers

What are the volunteers' motivators to make such big private investments without a tangible payoff? The following list may not be complete, but these are answers that one hears as responses when one asks this question.

---

* (Shakespeare, 1599).
† 467,171 members by end of September 2015.

- **Having impact.** Some of our volunteers may have a major budget and head count to manage at work, but in relation to the size of the company that they work for, their work's impact is still nevertheless small. In their volunteer work, they create results that matter.

- **Contributing to a great team.** The joint work of the team members brings about results that a single person would not be able to produce and helps forge strong networks based on common experience of success. There is no business case for networking, but strong networks can help master difficult moments as they regularly occur in any professional life.

- **Limitable time requirements.** Other tasks may come unexpectedly in their way, and volunteers also have to spend more time for their families, jobs, and other commitments. It is important to have mechanisms in place for such moments by having contingency plans working around their unavailability, or people who can take over their volunteer tasks and replace them. If people who are asked if they want to take over a chapter task feel that they cannot dedicate sufficient free time to volunteering, they will probably not do it at all.

- **Visibility in the community of colleagues and relevant organizations.** It is important that the names and faces of the volunteers are communicated, and that the organization's successes are publically associated with them.

- **Learning experience.** Volunteers gain experience and knowledge that they may not be able to obtain at their paid workplace. This may give them self-confidence, if they want to apply for jobs in the future that require such skills, this allows them to honestly document their volunteer work and its results in their resume.

- **Endorsement letters based on performance in a job.** Endorsements are a great tool. Professionally written, they recommend both the endorsed person and the person writing the endorsements. Volunteers use these volunteer endorsement letters as they use reference letters from previous employers. The volunteer endorsement letters often illustrate that the volunteers mostly do the tasks that they enjoy doing, so the recommendation is to use them for such tasks. Reference letters from employers are common in Europe—in Germany for instance, an employer must write such a letter in a benevolent style for a leaving employee. In countries where these letters are rather uncommon, an endorsement letter may be even more valuable, because it may be the only third-party assessment of the person's working style in a real project.

- **From time to time a party.** Events that allow volunteers to join and share the perceptions of impact, contribution, learning, and group experience are important to volunteer organizations.

This chapter has also shown the importance of the time aspect of the work. Team spirit is comparable to a jar of heavy ceramics that arduously must be carried up a mountain. Once dropped and broken into pieces, one may try to glue the shards back together, but it will no longer be the original jar, the cracks will remain visible, and it will be probably not be able to hold any liquid. With every difficulty that has been mastered, with every problem that has been resolved, individuals and organizations that have teamed up for the tasks may develop trust in the mutual reliability of the parties involved and in the endurance of the team spirit developed. When people feel fooled and exploited, they often turn to non-cooperative behavior too, and will make others in the project turn to the same behavior. Mutual behavior that ends in a dilemma situation is detrimental for the project as well as for many of the people involved.

Trust building is not a one-time activity, but needs continuous re-enforcement, and one should also not forget Stephen Covey's statement: "If you want to be trusted, be trustworthy"[*].

In 1984, Robert Axelrod developed a discipline called "Cooperation Theory" by expanding game theory with a simple question: Humans cooperate and have done so at all times. What helps them overcome the dilemmas of game theory and the Nash Equilibria that we find so often? His answer is that it does not take trust or friendship but durability of relationships[†]. It seems important that while our projects may be temporary and rather transient, our relations should not be that. The most successful strategy for a player in his research on "tournaments", repetitive game-theoretical situations, was the strategy of a nice person, who was never the first to defect, but immediately returned defection with defection, cooperation with cooperation. The player should never envy what the other players get or try to excel them, but instead develop care about the welfare of others, and the player should never cheat. This is precisely how people carry heavy loads and bring projects to a successful end, and it is a strategy that brings joy and satisfaction into people's lives.

Another important aspect is individual and collaborative motivation. This is a difficult task. In my classes, I often ask my students what food they like best. Some prefer a steak or grilled sausages. Many project managers from India prefer spicy vegetarian food, people from Arabic countries often like meat, but, being Muslims, do not eat pork. The reasons for these preferences or choices are different; however, it means that if the way to a person's heart goes through the stomach, one has to know the person well in order to use the way. Interestingly, hunger, I mean true hunger after five days without food, is the same for everyone. If we want to motivate people by helping them be satisfied in a time of hunger, a dish with bread or rice may be perfectly sufficient for any of them. If we want to motivate satisfied people by offering them the meals they like, the meals that bring them joy and delight, we must know their desires and what they reject. A common reason for game-theoretical dilemma situations that I observed in projects are managers on various levels, including project managers, who did not invest sufficient time and energy to know their stakeholders. Then, they do not know what it takes to motivate individual team members to place the team and the mission success before their individual interests and grow to become a Band of Brothers and Sisters.

In Chapters 3 and 4, I will present a rough typology of projects, discuss practices that are situationally favorable or detrimental for each of them, and discuss some methodological tools that I found helpful for Situational Project Management (SitPM). I will describe this considering the contents of this chapter: The organizational backgrounds and the game-theoretical considerations, both in essence being descriptions of the dynamics of success and failure in team situations.

---

[*] (Covey, 2004, p. 43).
[†] (Axelrod, 2006).

Chapter 3

A Typology of Projects

Russell Archibald[*] correctly stated, in 2013, that a "globally agreed" typology (in his words: a project categorization system) is "urgently needed". He further said that the absence of a typology often leads project managers to apply inappropriate "one size fits all" methods and that projects fail because of the lack of appropriate methods.

Project management is currently in a situation similar to chemistry before the discovery of the Periodic table in 1860, when Chemistry and Alchemy were often mixed. Biology was changed in a similar way when Linné developed his taxonomy in 1735. Before this moment, asking a student which species or genus he or she would want to relate his or her studies to would have been meaningless: species, genera, families, and so on had not yet been identified. In addition, the mechanisms of mutation and selection that allow species or genera to evolve over time hadn't been identified yet. In 2014/2015, I did some research, responding to Archibald's statement, to better understand which types of projects exist and how a project manager should adjust practices to them. The work had a focus on Germany, but the results are probably not culture driven and can be translated into other countries as well. Many of the results were as expected, and confirm the approaches and behaviors that one would expect from experienced project managers with a situational set of attitudes, but some results came as interesting surprises.

3.1 Introductory Questions

The introductory questions are intended to test your knowledge on certain project types and on the practices that can be beneficial or detrimental for them. They will all be discussed in this chapter, so if you do not understand a term or the concept with the term, come back when you finished reading the chapter and read the question again.

---

[*] (Archibald, 2013, p. 10).

1. A company is running its projects "over the fence". Which practice does this describe?

   a) Each project is performed along a predefined sequence of business units, and each phase is done by another unit. There is no project manager supervising the entire life cycle.
   b) Several business units work concurrently on a project, and each department works in tight cooperation with the other departments. A project manager administers the cooperation of the units.
   c) Several business units work concurrently on a project, and each department works in tight cooperation with the other departments. The units organize the cooperation among themselves.
   d) Each project is done inside just one performing business unit, which provides all management functions and the necessary resources from initiation to formal closure.

2. You are performing a project with a highly experienced team that has done a number of similar projects in the past. Your staff members understand their business and so do you. What basic mindset, as the project manager, should be most favorable for your project?

   a) The low degree of novelty for your team members will make your project easier and simpler and will generally reduce the risk of the project team to run into troubles.
   b) One can generally reduce contingency reserves put in place to protect deadlines and the budget and rely on the predictability that comes from the repetition of earlier projects.
   c) This is a perfect project for a part-time project manager. The team members will be self-organized easily and use their proven competency to resolve problems whenever they occur.
   d) One should avoid an attitude of easygoingness and complacency that may come with the repetitiveness of the task and make the team inattentive to the specific risks of the project.

3. You have been assigned as the project manager to a customer project under a fixed-price contract to develop and install complex digital equipment for eight operation rooms in a new hospital. The contract includes major liquidated damages for delayed handover of the equipment. The customer is a government agency that also performs the construction project for the new hospital, which is currently in its planning and approval stage. The actual construction of the hospital will begin in the next few months. In this situation, what are your major concerns?

   a) Focus on your own project: The hospital project is not your business, but you have to make sure that your deliverables and installation teams will be available on time as has been contractually agreed.
   b) Reduce documentation to a minimum: The complexity of the clinic construction may lead to delayed provision of data that your project needs, and you do not want to cause problems for its project manager in such cases.
   c) Observe the progress of the principal project: This phase is highly critical, as approval stages in public projects are a common source for major delays that in turn can impact the technical and financial success of your project.
   d) Focus on avoiding outlays for the public project: The hospital will be finished late anyway. Add 25% to the construction time and book your resources for the installation time that results from this calculation.

4. You identified that your project is a typical "Brownfield project". As a project manager, what is a practice to avoid specifically for a project of this type?

   a) You should take particular care of legacy solution that are still operational and may influence the performance of the project and its deliverables.
   b) You will have to identify many different stakeholders, understand their feelings, wishes, and needs, and adjust the project accordingly.
   c) You should focus on the technical details of your solution and not get sidetracked by interests of people, who are external to the project.
   d) You should expect frequent change requests from various stakeholders and establish a change request management system in place to control them.

5. For your new association project, a number of volunteers have approached you. They are prepared to invest energy and private time and desire to contribute to the project with the knowledge and skills that they have. Some have aspiration to work on new tasks and gain new experience and knowledge. How should you plan work packages and activities for the project?

   a) Build a classical Work Breakdown Structure (WBS) by decomposing the project into smaller, more manageable elements, and assign the team members as workers to them.
   b) Rely on the ability of the volunteers to become self-organized and resolve all problems for you and just leave the team alone. If every volunteer does his or her job right, the project will automatically be successful.
   c) Develop a project schedule using project management software and plan work items for the volunteers; and then send messages to the volunteers explaining the work items and your expectations on completion dates.
   d) Discuss with each volunteer what work items he or she wishes to work on and considers valuable for the project. Then coordinate the work of the team members in a way that all work will be done.

6. You are managing a customer project to develop and implement software on a fixed-price contract, and you do not think it will make a profit. You are following agile principles, and requirements and design are developed concurrently with the solution. Applying the Agile manifesto ("Responding to change over following a plan", "Customer collaboration over contract negotiation"*), the customer adds new requirements once a week but rejects renegotiation of the price or limiting the scope of your contract. What should you do?

   a) Meet the responsible customer manager in a neutral environment and discuss the one-sidedness of the contract implementation. Find a new agreement that protects your company from customer-side uncertainties.
   b) Talk with the customer-side employees. Make them understand your problems and ask them to avoid any communications that can lead to new requirements, allowing your team to develop against existing requirements.
   c) Some people say that it takes 20% of the time and effort to meet 80% of the requirements. Identify these 80% that are easily met and tell your team to ignore the other 20% of the requirements.
   d) As long as the customer is not able to provide clear and stable specifications that you can rely on to develop the software, you should stop the development work and shift resources to more profitable projects.

* (Beck, K., et al., 2001).

3.2 Best Practice Approaches vs. SitPM

On March 28, 2008, I had an interesting meeting at London Heathrow Airport. It was interesting in that it was the day after the "grand opening" of the new Terminal 5 (often just called T5) at Heathrow—a key moment for the two organizations, British Airways and British Airport Authorities, who ran the project jointly and also operated the terminal together. Except that the terminal was not operating by that time at all; instead, it was a huge disaster. Just as the operational opening had been started the day before, the systems in the terminal crashed, and the entire operations of British Airways fell apart. At one point in time, the company was unable to fly any aircraft to and from the terminal, and it took the company months to return to perfectly normal operations. I had a flight booked to Heathrow with British Airways the next morning. Late the evening before I was scheduled to depart, I found out that my departing flight, as well as the return flight scheduled for the same day, would not take place at all. I booked the last seat on a Lufthansa flight to Heathrow and back. I will discuss the lessons that I learned from this experience later in the chapter.

The meeting was appointed with a person who I will call Wendy in this case story. Wendy was a manager in a British training company that was interested in my support as a trainer for them. The topic was not project management, but bid and proposal management, something tightly linked with project management: During bid and proposal management, many decisions are made that can later make the lives of project managers easy or difficult. Proposal management is a discipline with specific professionalization and its own body of knowledge, a discipline in which the need for training is high, but the demand on the market is astonishingly low. I am a certified trainer for this discipline by the Association of Proposal Management Professionals (APMP), but I do not do much training business in this field. Most companies expect their bid and proposal managers to just learn by doing, and they are much more prepared to invest in their project managers. The British training company seemed capable to overcome this hurdle and turn the need for training into actual business. Commercial matters between the company and me were determined, and the meeting then was to discuss seminar materials, structure, and contents. I had studied the book that was mostly considered the body of knowledge of the discipline and would be part of the materials that students would receive. I had found some minor flaws, no big errors, but nevertheless statements that contradicted more recent research (such as, "serif fonts are more readable then non-serif fonts") and should not be used. I told Wendy about them, her response was: "Oliver, don't touch it. This material has been proven repetitively, and it works. It is 'best practice'. Don't alter anything". I think I never heard the expression "best practice" more often than in the two and a half hours that we spent together in the meeting room at Heathrow Airport. It was a perfect argument to reject any criticism and kill all opportunities for improvement. A good practice may be rethought from time to time, or critically assessed if it actually is good practice in a given situation. And there is always the option open for a better practice. In contrast, as soon as a "best practice" has been identified, the search for the right approach to a given situation is finished. Period. The uniqueness of projects is then no longer considered, nor, in business development situations, is the uniqueness of the customer organization with its specific desires, needs, and requirements, possibly fear and concerns, which the bid or proposal needs to address in its contents and in its tone. Developing a complex business is a project in itself, and it is subject to the same dynamics of success and failure as any other project.

We had some great classes, after the meeting together, and I also took opportunities to help companies in a consulting role to win business. The little adjustment that I made in class was to ask students to consider the practices "good practices for many business development and proposal writing situations" and to deliberate for each situation first, whether it belongs to this group or not. This means that one first needs to know the prospect, and I believe that this should be the first rule for any business development project. The joint business was not as intensive as we hoped. As mentioned earlier, many companies invest a lot in their project managers, but regard bid/proposal management a common sense discipline that can be done by any person with the right templates and therefore needs no special instruction. I think the assumption is wrong and the impact on business is too high to leave this discipline to trial and error. A company should hire or develop professional experts for their bid/proposal development to make sure that they win the attractive customer projects and keep their competitors busy with poor business.

So, what is a best practice? According to Stuart Bretschneider, Frederick J. Marc-Aurele, Jr., and Jiannan Wu[*], to call a practice "best", it must have three basic characteristics:

1. A comparative process (with other practices)
2. An action
3. A link between the action and some outcome or goal

There are many sources for "best practices" in project management that are considered universally applicable procedures and techniques for the discipline that a project manager can simply learn, adopt, and apply. An example is a project management methodology called PRojects IN Controlled Environments (PRINCE2), which is widely promoted to practitioners as a "proven best practice"[†] and can be "used successfully for all types of projects". Well-known and respected book author and speaker Harold Kerzner linked "management best practices" with "achieving global excellence"[‡]. Tom Mochal promises "10 best practices for successful project management"[§]. If one looks at the three criteria enlisted by Bretschneider and his colleagues, only the second is met by these promoters: The action. A comparative process, such as "We tried three practices in three perfectly identical projects, and found the first process to be the best" is nowhere described, and a clear link between the practice and the result is also not recognizable. Another open question is the dimension to which the superlative "best" refers. Is it monetary, strategic, social, or something else? A process that focuses on strategic elements may be detrimental on the monetary side and vice versa. Similar to a drug, each approach has its benefits and its unwanted side effects. Depending on the benefits that a person or an organization wants, and the side effects that it is prepared to accept, the definition of "good" may differ and with it the definition of the superlative "best". While project objectives are different from project to project, so is the project environment, and practices must be assessed in the context of these objectives and the environment. With this situational viewpoint, one notices that a practice that has been found "good" in one moment may be detrimental in another one. This

---

[*] (Bretschneider, Marc-Aurele, Jr., and Wu, 2005).
[†] (Axelos, 2014).
[‡] (Kerzner, 2010).
[§] (Mochal, 2009).

is the fundamental principle of what I call Situational Project Management, or in short SitPM. It seems obvious and self-evident to me, but what do others think?

The term "best practice" has gained high relevance in the world of project management practitioners; it is used to sell methodologies, books, services, training, software, and also conference tickets to project managers. Best practices are sold similarly to cooking recipes with the promise that a beginner only needs to learn the recipes and can then work at the same level with the most seasoned colleagues. They are also used to cajole customers. If a consultant writes a book on the "best practices" for his or her customers, the person stabilizes the business with these customers. The book will still be on the market place and tell the same story when the business environment of these customers has changed, and the former "best practices" are no longer appropriate. The book then may be a best-seller, because business people love success stories, assuming that they can easily copy successful organizations and be also successful[*]. The fact that the same approaches that have been successful at one time, in one situation, may lead into disaster in another is commonly ignored. This can happen very quickly. Many readers will remember Nokia cellphones. Here are some data on them:

Share prices (New York Stock Exchange, Closing prices, Yahoo-Finance, n.d.[†]):

- 6-11-2002: $17.53
- 6-11-2007: $41.10 (+134.4%)
- 6-11-2012: $2.77 (−93.3%)

Brand value:

- Millward Brown Optimor rated Nokia's global brand value in 2008 among the top ten. In 2012, they no longer rated Nokia on their list of the global top-100 brands[‡].
- From 2007 to 2009, Interbrand ranked Nokia fifth in the world. They rated Nokia 19th place in 2012 and 57th place in 2013[§].

Market share[¶]:

- 4th quarter 2007: 50.9%
- 4th quarter 2012: 2.9%

Profits before tax (million Euros)[**]:

- 2002: 4,917
- 2007: 8,268
- 2012: -2,644 (loss)

---

[*] A great recommendation on the topic is Philip M. Rosenzweig's *The Halo Effect . . . and the Eight Other Business Delusions That Deceive Managers,* in which he describes nine common delusions for business writers seeking to explain successes and failures and for company managers trying to copy success stories. Replace "companies" in his book with "projects", and "managers" with "project managers", and one finds a valid description of a lot of project management literature (Rosenzweig, 2007).

[†] (Yahoo-Finance, n.d.).

[‡] (Millward Brown Optimor, 2008), (Millward Brown Optimor, 2012).

[§] (Interbrand, 2013).

[¶] (Statista, 2013).

[**] (Nokia, 2003), (Nokia, 2008), (Nokia, 2013).

In the five years from 2002 to 2007, the Finnish company more than doubled its shareholder value and was mentioned among the ten most valuable brands in 2007, a year in which every second they made a cellphone sold in the world. Nokia's practices where highly successful during that time. Then, in the next five years, from 2007 to 2012, the share value dropped to under ten percent of the 2007 peak value; it not only became marginalized, its brand value had also dropped dramatically, and its profits had turned losses. The Nokia cellphone business was finally taken over by Microsoft in 2013. Nokia generally applied the same practices during 2002 to 2012, but the practices that led to success in the first five years caused failure in the second. What happened in 2007 that changed the business world so dramatically for Nokia? It was the emergence of two new competitive frontiers:

- Chipsets that allowed small manufacturers to quickly and cheaply develop new cell phone models and bring them to the market place. This business model was driven by the Taiwanese company MediaTek; it was a challenge to Nokia's business at the inexpensive end of its product range.
- Smartphones: Apple's iPhone and in its wake the Google-Android "ecosystem" with companies such as Samsung and some smaller companies mostly from Korea and Taiwan. These vendors assaulted Nokia at the high end.

In an article from 2006*, two employees of Nokia's network department described how the company wanted to use what they called "supply chain agility" to develop systems together with their suppliers that could respond quickly to changing demand situations and reduce procurement costs. Nokia had a major reorganization in 2007 and focused on improving production and distribution of its existing products. In 2007, Nokia also closed its operations in Bochum, Germany, and moved the production facilities to Cluj, Romania, where the company intended to develop a "Nokia Village" together with its component suppliers to profit from short delivery distances and just-in-time ordering. Nokia had an operating system and the basic software to control the electronic functions, called Symbian; using Symbian widely across its program was considered another way to reduce costs through scaling effects. These developments can be considered an appropriate response to the cost challenge that came from MediaTek. It was the kind of practice that Nokia had implemented over the last years and that made Nokia the market leader.

Nokia's slogan was "connecting people". I had some opportunities to talk with Nokia personnel in 2007, and they told me what they made out of the slogan: "disconnecting families". This mainly referred to the many hours of overtime work that Nokia staff was expected to do, but also to the secrecy that was expected from them, which made it impossible for them to talk about their work at home. It further referred to the frequent re-organizations at the company that left it unclear for many employees where their place was inside the organization, something that caused workplace stress, an anger that many employees vented at home. Nokia's business approach was always that of an introspective closed skill: If the company did its job better than others, it would be more competitive, and customers would prefer its products. As discussed in Chapter 1, Nokia as a company viewed itself as the figure-skater, and the market

---

* (Collin & Lorentzin, 2006). Note: I believe the authors confuse "Lean" and "Agile" concepts. The article deals with a lean approach, addressing uncertainties and fluctuations in demand, not an agile one, which would focus on uncertainties in functional customer requirements.

was the jury. The better its performance, the more acclaim and sales it would have. Nokia was not prepared for competitors such as Apple and Google, who did not try to make things better but different and in tight interaction with their business environments, which became an essential co-player in their performance. They understood business as an open-skilled discipline. In 2007, the first iPhone was launched, as well as Google's Android operating system for smartphones; the first smartphones based on it followed in 2008. How did Nokia respond to the new competition? Their chief strategist Anssi Vanjoki, the company's second-highest manager, told the German newspaper *Handelsblatt* in 2009: "Apple has originally also caused a lot of sensation with the Mac, but they remained nevertheless a niche vendor. This will be the same with mobile phones". He also said, "In the past we have too much focused on the technical basics instead of optimizing the design of our phones. But I am trustful that we know now, what we need to do: Build phones that are easy to use and good looking"*. He was still thinking in technical product dimensions not in user experience and in "ecosystems". The big innovation was the capacitive screen, which allowed "swiping" gestures and a new user experience, while Nokia still used little touchpad keyboards. The change needed for Nokia was huge: Symbian was not developed for the new screen technology, but Nokia wanted to stick with it.

Apple and Google's other innovation was to open their operating system for third-party software, small programs called "apps", which turned the phone into a small pocket computer useful for a wide range of tasks. App developers became thus a part of the "ecosystems" that both companies planted, understanding their approach more comparable to the work of a gardener than of an engineer. While Nokia still engineered its products to be inexpensive and to connect people, an approach that was successful until 2007, Apple and Google were busy connecting themselves with the people, the users, and the developers, and then connecting these people with the world. Users were prepared to pay a high price for this connectivity while developers invested time to develop the apps that made the other part of the eco-systems.

The rollercoaster case story of Nokia shows how a practice that led to success in one moment may cause failure in another one. It shows how managers may think they have found the philosopher's stone, so that they can do without the philosopher. In the case of Nokia, the story spans ten years: five years of growth followed by five years of decline. In our project management world, the rollercoasters may run much faster; we may experience such dramatic changes in our business settings in ten months, possibly less. Nevertheless, many people in project management believe in our version of the philosopher's stone, the "best practice", and expect it to turn inexpensive lead into gold.

I wanted to know how popular the term (and the promise that it communicates) actually is. Between April and August 2015, I invited project managers from a global LinkedIn community to respond to a small survey to find out how popular the concept of "best practices" is. The project managers were asked to select the statement that they considered the best from the following list:

- "There are proven 'best practices' in project management. Applying them ensures repeatable success in any given project and project situation."
- "There are no 'best practices' in project management. The same practice that has been successful in one project or project situation may fail in another one, and vice versa."

---

* (Louven, 2009).

- "Based on my experience and observations, I cannot support either of the claims."
- "Other (please specify)", followed by a text field for comments.

The participants of the survey were asked to select the response based on their experience and observations as practitioners and/or experts in project management. The survey was answered by 189 individuals. The responses are shown in Figure 3-1.

The selection "Other" had a comment field, where alternative responses could be given. Most pointed in the direction of the first answer that "best practices" exist in project management, but questioned their repeatability and universality in statements such as:

- "There are best practices that increase the probability of repeatable success, but success cannot be ensured."
- "There are proven best practices, but in countries like mine, with a lot of informality, it's difficult to apply them, but not impossible. I think we should continue working on it."
- "Best practices exist ... depends upon the ability and maturity of the PM."

The participants were also asked to select their gender and their relation to project management with the options:

- I am a practitioner in project management (46.6%).
- I am a practicing expert in project management (19.0%).
- I am a non-practicing field-expert in project management (Trainer, Coach, Consultant, Auditor, etc., 8.6%).

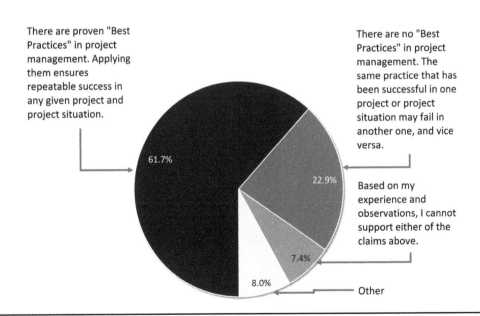

Figure 3-1   Responses to the options.

- I am a non-practicing off-field expert in project management (Researcher, Academic instructor, Developer of PM Software, etc., 2.3%).
- I am a beginner and/or assistant in project management and do not have much experience (20.1%).
- Other (please specify, 3.5%).

Figure 3-2 shows how the different group responded to the question. There was no statistically relevant difference between male (70.9%) and female (29.1%) respondents. With growing expertise, the skepticism on best practices seem to grow, but one should be careful with the off-field experts, as a number of four respondents is too small to be statistically significant. It is also that beginners and project assistants seem to be more skeptical towards best practice concepts; they may learn to believe in them in their work with project managers.

The respondents came from 49 different countries. I also wondered if there is a cultural bias influencing the belief in "best practices". For countries or regions with 10 respondents or more, I computed the percentages of positive answers as shown in Figure 3-3.

Interpret these numbers with care. There were between 10 respondents (Latin America including Mexico) and 54 (U.S. and Canada) who took part in the survey. These may not be enough to draw strong conclusions, and the method was not 100% scientific, using an online survey with self-selected participants. What the survey did show is that a major percentage of project managers believe in "best practices", and the popularity of this belief is stronger in some cultures than in others. Additional education will be necessary to help project managers develop a more critical and situational understanding of their discipline, especially given the large number of new project managers who are thrown into projects without proper preparation, education, and authorization, ignoring the disastrous effects on their self-esteem and their resumes.

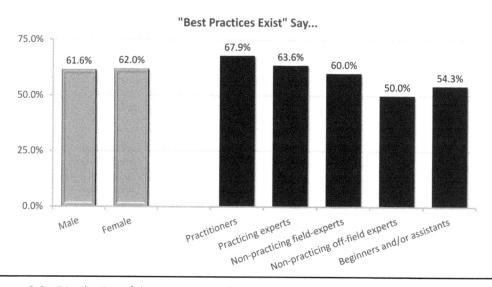

Figure 3-2 Distribution of the percentage of responses to the answering option "Best practices in project management exist . . .".

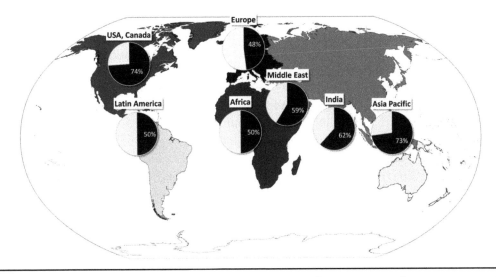

Figure 3-3   Geographical distribution of the percentage of responses to the answering option "Best practices in project management exist . . .".

3.3 A Research Project

Project management is a difficult discipline. The interaction of projects with their organizational, social, economic, and environmental context is highly complex. While project managers are active agents of change by administering investments that increase the adaptability of organizations and their ability to innovate, at the same time, projects and the environments in which they are performed are also subject to change. Many requirements, objectives, and constraints that were valid at the onset of the project may not be valid when the project comes to a close. This is because business situations and interests, management strategies, standards, and laws may have changed during the project lifecycle. Another aspect of complexity is the continuous need for interaction with stakeholders: people, groups of people, and organizations who are potentially influenced by the project, intentionally or not, and may in turn influence the project[*]. Stakeholders are those who need attention and who can claim changes on the project and its outcomes, vested at times with good cause, not vested at other times. While project management is often considered an engineering discipline, managing stakeholder perceptions and engagement are typically not considered an engineer's skill because of their generally non-quantitative nature. In 2014/2015, I did some research with the University of Liverpool, U.K., to address some specific questions that arise in this highly organic, dynamic, and discontinuous field[†]. In this chapter and in Chapter 4, I would like to report the results of this research and draw conclusions for our practical work.

The most characterizing attribute for a project is "unique", as discussed in Chapter 1. The uniqueness of projects sets them apart from operations, whose trademark is repeatability.

---

[*] (PMI, 2013, pp. 30–33).
[†] (Lehmann, 2015).

Responding to this uniqueness is what sets SitPM apart from the "one size fits all" concepts that are often communicated in literature, but easily identifiable by their ignorance of the simple fact that the same practice that has been successful in one project situation can fail in another one. Applying situational intelligence is a more difficult task, because one has to replace book smartness with street smartness and simple, quick solutions with considerate awareness of the consequences that actions can effect, intended results as much as unintended ones.

Uniqueness is not that difficult to manage and to understand. A common way to help structure unique things is to classify their variability by using typification. Skin burns, for example, are each unique as well as the things that caused them. Different burns need different treatments; otherwise, the treatment may not be helpful and may even increase the damage of the skin. In medicine, a typology has been developed for burns with several degrees[*]:

- First-degree burn: The outer layer of the skin (epidermis) is dry, red and painful.
- Second-degree burn, partial thickness: Affects the epidermis and upper parts of the dermis.
- Second-degree burn, full thickness: Affects the epidermis and most parts of the dermis.
- Third-degree burn: Destruction of all layers of the skin and subcutaneous tissues.
- There is even a fourth degree mentioned sometimes that involves muscles and bones.

For each degree, proper treatment is different. A first-degree treatment may be done at home (be careful with children!), and after a week, the burn should be healed. If one has a deep, second-degree or worse burn, one may need to seek clinical treatment. One can also classify burns according to their cause, for instance open fire or electricity, or measure the surface area affected. While each burn is unique, typification and classification helps determine the appropriate treatment. Classification can also help in the ongoing treatment of the burn. Specific research and development has been performed for the different degrees, and the findings of the research for a specific degree are not thoughtlessly carried over to treat burns of other degrees.

In the introduction of this chapter, I compared project management to Chemistry before the invention of the periodic table, its most fundamental classification system. Chemistry by that time was rather alchemy. Although project management is much less a science than it is a very practical task, we should require evidence for claims that are made and distrust promises of simple solutions when the task of project management is a difficult and complex one. Projects often must be navigated through unchartered and unmapped waters, of whose existence the promoters of the solutions were not aware.

3.3.1 The First Objective: Develop a Typology

My research had the intention to develop an open typology by talking with field experts and using their anecdotic memories to identify underlying structures. Field experts are people who are not directly involved in managing projects, at least not at the time of the research, but work closely with project managers yet see projects from a sufficient distance to identify structures,

---

[*] The example is given as a model and for reference purposes only. Please always consult a doctor or a hospital for the correct classification and appropriate treatment of real burns.

trends, and commonalities that an insider may not see that easily, especially given the degree of passion and emotion that is part of the profession of many project managers. Some things are better examined from inside, by the person who has gathered the experience, others by an outside observer, and I was interested in the second. I could finally forge a group of 17 field experts, among them instructors, book and magazine authors, PMO managers, and volunteers in a project management association (PMI®); the latter were active in two roles. They had between seven and 36 years of experience in project management, the average being 23 years. Sixteen of the experts were German, one was Swiss but living near the border to Germany and did a major part of his instructing business in Germany. Twelve of the experts held the PMP® certification*, one held in addition the IPMA Level B certification†. The experts came from diverse industries, which included automotive, consulting, education, electronics, information technology, and publishing. One may argue that the concentration on experiences and observations of a group of persons may be too subjective as a scientific approach, but given the complexity of the relationship between project managers, their projects and the environments, in which the projects are performed, the approach seemed appropriate to me, and a cumulated time of 393 years cross-industry experience by persons in mostly senior positions is a great source of rich knowledge to be tapped. Objectivist future research may confirm or reject the findings I made.

To ensure the integrity of the research and avoid bias, the selection of field experts excluded persons who met the requirements defined but had a direct association or business relationship with me.

The first objective of the research was to develop an open typology. "Open" means that the dimensions identified are not considered complete. While my research pointed to the existence of certain typological dimensions, it definitively missed others. I am personally aware of some more and will describe them later in this section, and others may know of dimensions that I have not identified so far.

I considered it important to have a well-defined process for the research to allow others to repeat it and confirm what my experts found or find different results. The few existing approaches to develop project typologies do not describe the development process and leave it open whether their results were achieved following clearly described and implemented methods or were just brilliant ideas that someone got in the bath tub. The process that I used was a combination of several methods, each of which should have its place in the toolbox of every SitPM project manager:

1. **Critical Incident Technique.** A person is asked in an interview to remember a moment in his or her life that is connected with strong emotions and has been found important by the person. These incidents are the kind of memory that the human brain stores for a long time, while seemingly less relevant memories get forgotten. I adjusted the method slightly by asking each expert two questions: a dysfunctional one that calls

---

* Project Management Professional, a certification offered by PMI®, the Project Management Institute.
† The person's certificate was actually awarded by the Swiss Project Management Association (SPM), whose credentials are accepted and internationally harmonized under the umbrella of the International Project Management Association (IPMA). At IPMA, Level B is the second-highest level. It stands for Certified Senior Project Manager. IPMA and PMI® are the two global competitors in the market for professional certifications for project managers.

for discomforting memories and a functional one that asked for pleasant memories. Research on the processing of odors, music, and pictures has shown that pleasant and unpleasant memories are processed in different brain areas, and I assumed that the same is true for project management memories. The answers during the interviews in my opinion confirmed this assumption. In project management, the tool is helpful for lessons learned interviews and workshops, especially if they are done late and not immediately after the event.

2. **The Five-Whys Method***. This method is commonly linked with the Toyota Production System to research the root cause of an obvious problem by asking five times why: We have a problem, what it its cause? This cause is also a problem, why do we have this problem? And so on. The Five-Whys method is a powerful tool to find systematic but invisible determinants by assessing anecdotic but observable incidents. In project management, a common observation that I made was that technical problems and resistance by stakeholders had their root cause in team fragmentation and even fear.

3. **KJ Analysis.** Named after Kawakita Jiro and also called Affinity Diagramming; another of these great Japanese methods that help to find the invisible and systematic structures by assessing visible things. They seem to be trivial and simple at first glance, but when one applies them, they develop force and help to dig deeper into issues. We used it mainly to consolidate the dimensions identified. As an example, teams that are dispersed over geographical distances, cultures, and time zones, and others that were built from employees from different business units or different companies, basically had the same problem to cope with: Fragmentation. We called them "siloed projects" in contrast to "solid projects" in which no such fragmentation exists. Steps 2 and 3 were done in an online team environment that was also used for the fourth step.

4. The experts then received a brief introduction to two models for project practices, which will be discussed in the next chapter in detail:
   a) Lifecycle approaches between "Predictive" and "Adaptive" (three approaches)
   b) Lipman-Blumen Achieving Styles (nine styles)

5. In a next step, the experts were asked to respond to an online survey and answer for each type of project whether they found a specific life cycle approach favorable or detrimental, and the same question was asked for each achieving style. For the assessment, a seven-point scale was used with the extreme positions "+3 = highly favorable" and "–3 = highly detrimental" and a middle position "0 = makes no difference".

I will discuss steps 4 and 5 in Chapter 4 and also present the results from the survey there, which will then lead to recommendations for both favored and detrimental approaches and achieving styles. Here in Chapter 3, my intention is to give you an understanding of the identified types of projects, and Chapter 4 will then discuss which practices are most likely helpful to manage them.

For step 1, the interviews using the modified Critical Incident Technique, the experts were asked to answer the following questions:

---

* I would prefer to call it the *n*-Whys method. The focus on the number five is somewhat esoteric, and I have often used it successfully with fewer (but never with more) than five whys.

- **Dysfunctional question:** "During your time as a practitioner or as an expert, do you remember a moment when a practice, a method, a behavior or a tool for project management, that had led to success before, led to failure?"

- **Functional question:** "During your time as a practitioner or as an expert, do you remember a moment when a practice, a method, a behavior or a tool for project management, that had led to failure before, led to success?"

The approach worked as intended with 14 experts, who each remembered two different cases, and one expert even memorized three cases. Three experts could answer only the dysfunctional question but had no "pleasant" memory to share as a case for the research in response to the functional question. These were then further followed up through steps 2 and 3 to develop typological dimensions as shown in Table 3-1. B/W in the table stands for what scientists call

Table 3-1 The Typology of Projects and Project Situations

Typological Dimension		Mode
1	**Mark 1 project** First of its kind project for the project team.	**Mark _n_ project** Similar projects have been done in the past. Team members can build on experience and analogies, but may miss new influence factors that were not present in the past.	B/W
2	**Greenfield project** The project is performed in a "virgin" environment. The project manager can focus on technical matters and internal organization.	**Brownfield project** The project is performed in a developed environment. The project manager has to consider effects of the project on this environment and perform stakeholder management to ensure positive stakeholder support and engagement.	B/W
3	**Siloed project** The project team is dispersed across organizations, business units, locations, and/or time zones that lead to fragmentation of the team.	**Solid project** The project team is clustered in one organization, location, and time zone.	Grey-shades
4	**Blurred project** Work contents and objectives are unclear; various stakeholders have conflicting needs and expectations. The project may be a vehicle for political struggles.	**Focused project** The project develops and maintains a clear focus on its objectives and the way to achieve them and maintains it during its course.	Grey-shades
5	**High-impact project** The project brings massive change to the organization(s) involved and/or their environment.	**Low-impact project** The project brings low or no change to the organization(s) involved and/or their environment.	Grey-shades
6	**Customer project** The project is a profit center for the performing organization. The project manager creates the income for the organization and must ensure compliance with a contract.	**Internal project** The project is a cost center for the performing organization. Project managers must meet mandatory requirements or run the project according to a business case.	B/W

(continues on next page)

Typological Dimension		Mode
7 **Stand-alone project** The project progresses and succeeds in its own capacity, depending on the work that needs to be done and the availability and competencies of the persons involved.	**Satellite project** The project depends on another project ("principal project"), whose efforts are used by the satellite project. The approach may save cost and effort, but the project's progress and success depends on the progress and success of the principal project.	B/W
8 **Predictable project** The project allows for long-term predictions, resource booking, and work assignments, but may also require long-term predictions.	**Exploratory project** "The way is made by walking". Requirements and deliverables are not stated in advance by a requester or customer but must be identified by the project manager and the team during the course of the project.	Grey-shades
9 **Composed project** The project scope is developed individually by the team members and consolidated in a bottom-up way.	**Decomposed project** The project scope is defined and then decomposed and assigned to team members.	B/W

"dichotomies" in their language: black-and-white dimensions without grey-shades in between. "Grey-shade" stands for dimensions with continua between two extremes. I will describe these project types in the following chapters.

3.4 Mark 1 Projects and Mark *n* Projects

We took these terms from engineering and British sports car manufacturing. A Mark 1 version of a machine is the first generation, highly innovative, but also far from being optimized. A famous historical example is the British Ferranti Mark 1, the first fully electronic, general-purpose, and commercially available computer. Two units were built. The first was delivered to the University of Manchester in February, 1951. Interestingly, the next version was not called Mark 2 but Mark 1 Star, probably to further signal the still highly innovative character of its design; Ferranti sold nine computers of both versions. The designation Mark 1 is mostly avoided in order to not communicate the potential immaturity of a development to a market place. Between 1962 and 1991, U.S. auto maker Chevrolet numbered the follow-up generations of its first V8 (Big Block) engines used for powerful passenger cars and medium duty trucks Mark II to Mark IV. Japanese camera maker Canon did the same with its top-notch camera body EOS-1D, whose follow-up versions were called EOS-1D Mark II through Mark IV between 2004 and 2011. This allowed Canon to further use the established name but, at the same time, to communicate to the customer: We have an improved generation available. You should consider buying this product now, if you still have the old one. Mark 1 communicates revolution; Mark *n* signals evolution.

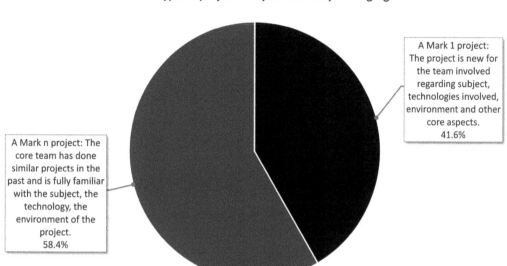

What type of project are you currently managing?

A Mark 1 project: The project is new for the team involved regarding subject, technologies involved, environment and other core aspects.
41.6%

A Mark n project: The core team has done similar projects in the past and is fully familiar with the subject, the technology, the environment of the project.
58.4%

Figure 3-4   Distribution of Mark 1 projects and Mark *n* projects among the respondents.

When I asked practitioners from a total group of 517 respondents, a majority said that their projects are of the Mark *n* type, but Mark 1 projects were not rare at all* (see Figure 3-4).

A Mark 1 project is new. It may be some great, major innovation, similar to the Ferranti Mark 1, whose development and manufacturing could build on the experience of its team to build experimental computers, but making a commercial computer was something new. Ferranti had to give customers the assurance that it would work without the flaws and safety hazards for users that people would accept in an experiment. The project must have been a challenging task for the developers. In other Mark 1 projects, the task as such may not be new at all but is new to the team. The novelty may then lie in the use of unknown technology for a known task in an environment that is yet unchartered, at least for the team or in anything else that distinguishes the new project from past projects to a very fundamental degree. I remember, for instance, the first one-day congress that the PMI München Chapter held in 2000. The chapter is a regional, not-for-profit organization incorporated under German law and works based on a charter agreement with the U.S. Project Management Institute (PMI®). The congress was organized by approximately 30 volunteers, two of whom acted as project managers for the event. All volunteers were seasoned project managers, but none of them had ever organized a congress. The original intention was to allow 10 months for planning and preparation. During initiation, members of the chapter board did a Novelty, Technology, Complexity, and Pace (NTCP) analysis as proposed in 2007 by Aaron Shenhar

---

* There was also an answer option, "I am currently not managing a project". The responses are not shown in this and the following diagrams.

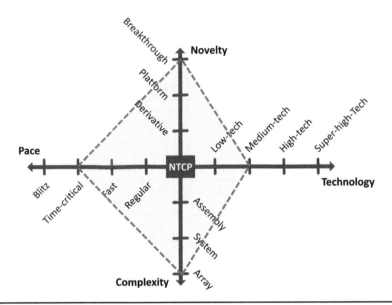

Figure 3-5 The NTCP model applied for a project in a not-for-profit organization. (*Source:* author's work, model used with permission of Aaron Shenhar. In a presentation made at the 2015 Global Congress of PMI, Aaron Shenhar, with Ori Orhof, presented an updated version of the diagram with four steps each for the Novelty and Complexity dimensions.)

and Dov Dvir*. As mentioned, NTCP can be interpreted as four typological dimensions. The project was rated in each of the dimensions along scales with three to four steps from low to high. The underlying assumption is that each dimension contributes to risk—the higher the project scores on it, the higher the overall risk is deemed. Figure 3-5 shows what the project looked like in the analysis.

- **Novelty.** A one-day congress is of course nothing fundamentally new. It was new to the participants, who were experienced in managing projects in IT, electronics, engineering, and other industries. None of the volunteers who performed the project for the chapter had any experience in conference planning and management. This lack of experience had several consequences: Decisions that an experienced event manager would do quickly took them more time for discussion and deliberation, and they also needed time to fix errors that they made, once they became obvious. The novelty also came with a disbelief in the group's ability to run the project. The manager of a company that I approached to support the project as a sponsor clearly told me that he had experiences with such events in the past and did not believe that the team will succeed with the task. He would not want his company logo linked with a failed event. Novelty was set at maximum: Breakthrough.

---

* The approach of developing an industry-independent project typology and recommending adaptations in management style is similar to my approach, but Shenhar and Dvir do not communicate the development process for the typology, which I consider important, and their results are described as a closed typology, assuming completeness, while I regard the results of my development work incomplete (Shenhar & Dvir, 2007, pp. 46–55).

- **Technology**. Some technology was needed before the congress for a call for papers and for a system to register attendees. Some more technology would be needed during the congress for sound and presentations, etc. This was easily available, so that we gave technology a medium-tech rating.

- **Complexity**. This referred more to organizational and interpersonal complexity than technological. The chapter's economic opportunities were limited by the time, and a failed congress could derail the chapter financially*. A congress needs many investments upfront, but most of the income for congress sponsors and tickets comes relatively late. Liquidity could become a problem. In addition, we had to take care of the interests of hundreds of stakeholders including our members, the long-term sponsors of the chapter, PMI as the chartering organization, and of course of our volunteers, who spent hundreds of hours of private and unpaid time for the project. Further complexity was on the marketplace where another association and some commercial providers run similar congresses, organized by experienced professionals, tapping the same budgets of potential attendants that we would need for our congress. We rated the project as "Array", the highest grade on the complexity scale.

- **Pace**. The original time frame from the initiation of the project to the date of the event was planned with 10 months, which would have been "time-critical" in NTCP language.

Discussions in the board and with the volunteers came out with an assessment that this approach would be too risky for the chapter. One solution would have been to hire a professional event manager, who would reduce the rating for novelty. This was something we did not want to do since it is part of the mission of the chapter to open opportunities to volunteers. The alternative solution was to give the team more time. The event date was postponed by nine months. Allowing the team almost twice the time made it easier for the volunteers to discuss open issues and make difficult decisions and allowed for more time to fix problems. Another interesting aspect was that it took pressure from the cash-flow side. It allowed for more time to find event sponsors and have income before early investments needed to be made, and sponsors would have the benefit from a longer period, during which they were visible on the congress website, so they could expect a higher return on their investment. As soon as the first two congress sponsors were on the website and showed their trust in us, it was much easier to find more sponsors to support our event.

When the date had been postponed, the NTCP diagram looked as shown in Figure 3-6.

The example shows how novelty can impact a project. In the chapter project, this impact could be made more manageable by allowing the team more time. In other projects that need to meet contractual or legal deadlines or have to deliver before a market window closes, such a solution may not be possible, and other solutions need to be found.

The development of the project terminology discussed in this chapter was based on experiences and observations of field experts, in which the various types occurred, and the topic, novelty, appeared seven times. It was in this research that we discovered that novelty was the most important factor that made the experts see practices fail that have been successful before and vice versa. This confirmed the novelty dimension of the work of Shenhar and Dvir. But there was also a surprising observation: Of the seven projects named by the experts, where novelty was

---

* With some successful congresses that the chapter has performed meanwhile, the financial solidity would be strong enough today to survive a failed congress.

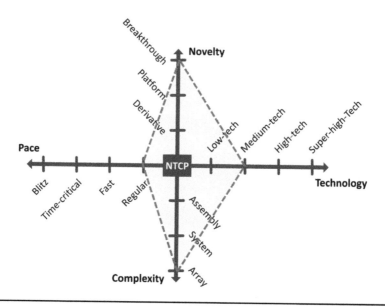

Figure 3-6   The revised NTCP model for the congress. (*Source:* author's work, applying the Shenhar-Dvir model, used with permission of Aaron Shenhar.)

the greatest problem, high novelty was only the problem in two of them. Five of them suffered from low novelty. The Mark *n* projects have risks that are greater than the risks that come from navigating in uncharted waters: routine, complacency, and ignorance of risk. With repetition, the aspect of uniqueness, the most characterizing aspect of projects, is more and more ignored, and a dulling feeling of routine creeps in, which was not expected. When such an attitude develops, one often hears statements such as "we have done this before", "we know how to do this", "we have a proven formula", and "we have developed excellence in this field". Then, projects are not regarded as new challenging endeavors in which stakeholders must remain alert for the unexpected, but as simple derivations from former projects. Repetition saves cost and time, as long as the project follows the plan. But then, seemingly marginal and negligible influence factors, which no one takes care of, can grow, cause major problems, and finally derail the project in an unanticipated way. I observed such problems frequently in software integration projects. The teams assumed that existing software in an organization was documented in a way that its process flows, data structures, and interfaces are easy to understand, and that interconnections between different software solutions can be built without difficulty. When the team then tried to connect its new solution with the existing ones, problems arose from incomplete, often false, documentation and from the many workarounds that were implemented in the old software solutions without any documentation, as hotfixes during implementation and operations, possibly years ago. For a small contractor working for a customer under fixed-price contract, such a situation can destroy the entire existence of the firm.

Another example that I came across was an international software rollout for a multi-national corporation in 28 countries, which went well in the first 27 countries, but stakeholders in the last country wreaked havoc on it. They had made themselves comfortable with special software that was developed specifically for them, and the fear of losing autonomy and independence was greater than the joy of getting a new and more modern solution. Part of the problem was probably that their old software had some functional deficiencies that allowed them to conceal operational inefficiencies and, as a consequence, to keep the work-force higher by numbers than what was necessary. Had this country been the first to implement the new solution, the project team would probably have been aware of the risks and would have addressed them in their approach; however, after 27 comparably easy implementations, the team just wanted to follow the so far efficacious formula and finish the project on time. Its members were not prepared for a new kind of problems with people. In the end, the corporation's upper management had to interfere, managers in the country operations got replaced, and branches lost autonomy in more fields than just software. For the project, it was this country that caused the team to miss schedule targets and to exceed a budget, which was originally based on realistic assumptions.

3.5 Greenfield Projects and Brownfield Projects

An interesting example is a comparison of two recent main-station mega projects in Germany: Berlin's Hauptbahnhof, after its completion in the year 2006, was mostly considered a major success, and Stuttgart Hauptbahnhof, whose renewal project ran into a state of deep crisis in 2010. It is interesting to compare the two projects (see Table 3-2).

A further commonality is that both projects were managed by the same project manager who generally applied the same practices using a highly detailed plan with a long planning horizon, which he rigorously protected from change requests in any form. What was the difference?

Berlin Hauptbahnhof was a Greenfield project in the literal meaning of the word. The fall of the Berlin Wall in 1989 and the reunification of the city of Berlin left a spacious but thinly occupied and mostly unused strip of fallow land crossing the entire city in the North-South direction. The wall was effectively a system of two walls with cleared land between them. It was often referred to as the *"Todesstreifen"* (Death Strip) because it was a complex fortification built to prevent the people from East Germany from fleeing to the West. Its demolition left a 43.7 km (27.3 miles) long and up to 500 m (550 yards) wide band of unused land, which gave city planners the rare freedom to develop a new town center without having to pay attention to people living nearby and for legacies. There, they could build the station and other infrastructure following the vision of the architect and the technical and organizational requirements of a 21st century traffic hub.

Table 3-2 Comparison of the Two Station Projects

	Berlin Hauptbahnhof	Stuttgart 21
Industry	Railway	Railway
Application area	Construction	Construction
Deliverable	Main station (new)	Main station (reconstructed)
Shareholders	Deutsche Bahn (German rail), local, regional, and national governments	Deutsche Bahn (German rail), local, regional, and national governments
Mostly regarded as	Success	In crisis (in 2010–2011)

Stuttgart Hauptbahnhof was different. The existing central station was constructed as a terminus, a dead-end station, where trains have to reverse directions. Termini are stations that must occupy much more space than through platforms, as the trains use the same track area to arrive at the station and to depart again. It is logistically more difficult to manage, and the travelling time lost for a train that needs to stop in the station on its way is much higher. Stuttgart is a major station on a currently developed high-speed railway connection from Paris, France, to Bratislava in Slovakia called "*Magistrale* for Europe", a program that is intended to cut the total travelling time over the full distance from currently about 11½ hours to eight hours. The railway construction in the Stuttgart area includes the partial demolition of the current historic (and iconic!) station building, building a new underground through platform station with eight tracks (replacing the 16 dead-end tracks) and 33 km (20 miles) of new tunnels built for speeds up to 250 km/h (156 mph). The problems began with citizens (stakeholders in the modern understanding of the word) questioning the consequences of the tunnel drilling for their homes on top of the underground construction site. The tunnels had to penetrate layers of anhydrite, which swells by 60% when in contact with water, and a lot of water is used for tunnel boring.

When geological layers swell, they pose a threat to buildings on top, and such a case had just happened not far from Stuttgart in a small town called Staufen im Breisgau, which had invested in a geothermal heating system for its town hall, for which engineers pressed water into layers of anhydrite under the historic town center. The swelling anhydrite affected cracks in foundations and walls of the houses above and threatened to damage gas pipes and other underground infrastructure[*]. The alarmed home owners in Stuttgart tried to contact the project manager to discuss their concerns, but he reportedly refused to speak with them. The plans had been open for public review some time ago, following applicable law, and the appeal period was over.

The ignored stakeholders then did what stakeholders commonly do in such a situation: they called friends and lawyers for support and looked together for vulnerabilities of the project, from hazards for endangered species to dangers for the ground water; major demonstrations followed, and the construction was massively disrupted. The following upheaval in Stuttgart and the state of Baden-Württemberg went far: The project manager resigned from the project, and the political party governing the state lost the next election. There were two periods of construction freeze until the work could be resumed. The estimated project costs increased during this time from 4.4 million Euros to 6.5 million, partially because of the freeze and changes to the plan that were necessary with the protests in Stuttgart. The project has started again and is expected to be finished in 2021.

The approach of the managers responsible for the project has changed dramatically since the time of crisis. Before 2010, there was a small Citizen Information Center with models and animation movies shown, and a simple website with some basic facts and news was offered for an interested audience. Since 2011, these media improved greatly. As a new medium, a public discussion platform was introduced called "*Bürgerforum*" (citizen forum), which allows citizens to raise their concerns during the project and discuss them with experts and representatives of the project shareholders. The objective is to ensure public support for the project and reduce damages and disruptions for the city to the minimum that is unavoidable.

---

[*] (Sass & Burbaum, 2010).

The shareholders of the project learned the hard and expensive way what it means to run a Brownfield project such as Stuttgart 21: the project team has to consider the influence of a large number of stakeholders, who cannot simply be ignored, as they (and their friends) often have options at hand to cause significant damage to the project. One has to manage legacies that can impact project success and has to limit the damages and operational disruptions that the project causes. In this type of project, the leeway for project activities is significantly reduced, and the need for intensive communications and rapport building are an additional challenge for the team.

I believe that the project team in Stuttgart is capable of constructing the tunnels without jeopardizing buildings on top. Stuttgart is surrounded by hills, and many tunnels have successfully been drilled there, so a lot of experience in tunneling through anhydrite layers is available. What the team was not able to do in 2010 was to build trust in this capability by people possibly impacted and engage them as supporting stakeholders.

Practices favorable for Greenfield projects may not be appropriate for Brownfield projects and vice versa. Does this principle also apply in other industries and application areas? A few year ago, I did training and workplace coaching for a number of project managers from a major U.S. company that implemented human resource management software in customer organizations. They had some successful projects in smaller companies that had no clear HR processes implemented, and the software projects were also business re-engineering for these customers: replacing existing ad hoc approaches with clearly described processes that were step-by-step supported by the software. These customer companies were comparable to Greenfields for the project teams and the software and could be performed with a focus on technical questions. Three of my students were then assigned to a project for a major government agency that had HR processes defined and in place, and the software was expected to support the existing processes rather than modifying the organizational processes to go with the software. My students were not prepared for the problems that came with this requirement. For them, it was just another Mark n project, repeating the well understood tasks of the previous projects. It is much easier to adjust an organization to a software program than vice versa.

The greatest resistance came from employees in the customer organization. They had been given formal job guarantees, so they did not have to fear unemployment or demotion; but they were not given guarantees that after the software implementation, they would still do the same job, remain in charge for the same tasks, and remain in their same offices, where many of them had been for years. As mentioned in Chapter 1, people personalize their offices and become attached to their environments. These people also felt they would lose the familiar job environment and were frightened of the unknown future for them. The project depended on their cooperation and support, but it did not happen. The agency employees had to describe to the team how their processes functioned, so that the team could mirror them in software, but these descriptions were not given, or if they were given, they were late and incomplete. The implementation of the software was delayed by months, and the company lost money because of many hours that the team worked for the customer that could not be billed and damage claims from the customer based on the contractual conditions for late deliveries.

When I was asked to help, the manager of the project managers privately let me know that he considered his people too soft. Given the situation, I would disagree: a tougher approach toward the customer's employees would have increased the resistance by the people, whose support the project needed and who were in a position to damage the project in a way that the blame would have fallen on the contractor, not on them. Spending time, energy, and empathy on building rapport and engaging the employees in an early stage would probably have been more successful. They were project engineers who focused on technical aspects, more than project managers who would do stakeholder management and engagement. Project engineers are the right people for Greenfield projects. For a Brownfield project, one needs project managers.

How literal should one regard the terms "Greenfield" and "Brownfield"? When Nokia (see the previous case story) decided to leave Bochum and move their production to Cluj in Romania, their basic aspiration was to build a "Nokia Village" there, with outlets of their major suppliers in the direct neighborhood, to be able to ramp production up and down to follow the fluctuations of the market demand for their handsets. Bochum, densely inhabited and more expensive, would not have allowed this operational strategy. The area selected for the Nokia Village in Cluj was a literal green field. What they missed was that Apple, and in their wake also Google, developed new business on much wider and, as we know meanwhile, greener fields. They called them "eco-systems", and there, they developed an unknown interconnectivity between hardware manufacturers, app developers, service providers, and customers; and with powerful processors and capacitive display technology that allowed "swiping" and other simple gestures, they revolutionized the perception by users that a smartphone is something "natural". If Nokia in 2007 had understood the metaphorical and emblematic interpretation of "Greenfield" and reacted accordingly, they might still dominate the mobile market today.

3.6 Siloed Projects and Solid Projects

Who loves telephone or video conferences? Without the perception of local proximity and physical presence, the ability to fully use body language and facial expressions and with common disruptions that are often due to unreliable line quality, the meeting experience is more uncomfortable than pleasant for most people—at least those coming from business. A common observation is that people hope to replace the development of true team spirit with technology, because the first is expensive, the latter is not. This meeting experience can be further deteriorated when the team is internationally dispersed. The project manager wants the project to move forward quickly, but instead it wobbles and the team spirit deteriorates over time. In addition, virtual projects are more likely to run into difficulty because of misunderstandings.

Here is another interesting case story: On December 11, 1998, the U.S. space agency National Aeronautics and Space Administration (NASA) launched a probe, called Mars Climate Orbiter, to the planet Mars. It was intended to study the atmosphere of the planet from a circular orbit around the planet and also to be used for the transmission of communications with another probe, called the Mars Polar Lander. That project began three weeks later. After a bit more than nine months, on September 23, 1999, NASA lost the Mars Climate Orbiter shortly before it arrived at its final location. As part of the final navigation, the probe was planned to "skim" temporarily through the upper layers of the Martian atmosphere twice at the height of 226 and 210 km (140 and 130 miles) over the Martian ground and use the drag of the very thin atmosphere there as "airbrakes". The probe instead dove deeply into the Martian atmosphere, down to an estimated 57 km (35 miles) distance from Mars's surface. The

Mars Climate Orbiter was designed to withstand the drag from Martian atmosphere down to a minimum height of 80 km (50 miles); at the 57 km height, the probe got destroyed or, if it survived, was then pitched into open space.

After the loss, NASA immediately created a "Mission Failure Mishap Investigation Board", which published a 35-page report on November 10, 1999, seven weeks after the incident[*]. The report said:

> Root Cause: Failure to use metric units in the coding of a ground software file, [called] "Small Forces", used in trajectory models.

The statement referred to a software program used to calculate the combined effect of airbrake force and brake rocket thrust. It passed its data to another software system, but numbers were false by a factor of 4.45, using lb-sec instead of the required and expected Newton-sec (1 pound = 4.45 Newton). Therefore, the rocket thrust was insufficient to navigate the probe into orbit as desired. The problem had existed over the entire course of the probe, and little errors in rocket firing added up to others, but remained unnoticed until the probe was lost. If it had been identified and communicated early, the problem would have been easy to fix.

The use of the term "root cause" at this point is probably not accurate. The next question in a "five whys" analysis would ask: "Confusion of metrics is a significant problem; what is its cause?" A simple answer would be that someone, possibly a subcontractor, did not adhere to a contractual specification, which required the use of the metric system. This answer would be a classic case of finger-pointing and may be useful in court, but would not help fix the problem and avoid repetition in the future, especially given the complexity of modern supply networks that require cooperative solutions. Another root cause could be identified in the concurrent use of imperial and metric units at NASA, following the Metric Conversion Act of 1975, which made it a policy to "coordinate and plan the increasing use of the metric system in the United States". Having two competing standards at a time inside an organization is a common cause for confusion and errors.

The report states some "contributing causes", which are also interesting:

- Navigation team was unfamiliar with the spacecraft.
- System engineering process did not adequately address transition from development to operations.
- Inadequate communications existed between project elements.

One may see these statements from an engineering viewpoint and implement processes to avoid them. This approach is what NASA obviously did. The result, however, was the development of a complex rule-book, too large and complicated to be fully read, understood, and implemented and leading to the opposite of its intentions. Also, it is not implausible that the requirement of using the metric system in the project was overlooked by people, because of the large number of requirements and rules that they have to remember and apply. Someone may have simply lost track of them.

I already mentioned Conway's law, stating that systems mirror their design teams. A communication slip-up between the components of a system has likely its root cause in a communication problem inside the design team. There were many organizations involved, including:

---

[*] (Stephenson et al., 1999).

- NASA, Washington D.C.
- Jet Propulsion Laboratory (JPL), Pasadena, California, a division of the Caltech (California Institute of Technology)
- Lockheed Martin Astronautics, Denver, Colorado, prime contractor
- Subcontractors over several tiers

The different players in the multi-player game had different interests. NASA as a government agency was accountable to politicians and finally to the public for accountable and successful use of tax money. Caltech was interested in the engineering and science aspects of the project (and probably also in the monetary income from it) to stabilize and further develop the position of the institute as a leading academic establishment. Lockheed Martin was interested in good news for its shareholders with profits and securing future business. Fragmentation had a geographical dimension but also a business aspect. While the silo structure of the project was probably unavoidable, the project needed to be managed in a way that addressed the issues that arose.

NASA by that time followed an approach called "faster, better, cheaper", which also had a negative influence on inter-team communications. Communications are costly and time-consuming, especially among geographically dispersed teams. An obvious and quick way to reduce costs and speed up an underfunded and understaffed project is to reduce the number and intensity of meetings and use team building, training, and other means of active networking. A basic problem of networking is that there is no business case for it. One cannot compute its expected monetary benefits against its costs and project a return on investment. The value of networking is only assessable from hindsight, when errors are made that are due to a lack of team communications and also a lack of visible team spirit. However, well-developed networks can help in coping with unexpected problems in an effective and efficient manner. I am certain that for Mars Climate Orbiter, the communication error of the software systems was preceded by a communication problem of the teams involved. In "faster, better, cheaper", the most urgently missing element is "people".

An objective of the investigation report of November 10, 1999 was to protect the concurrent and tightly related project Mars Polar Lander from similar mishaps. This probe was planned for landing on December 3, 1999, and it also got lost, together with two probes called "Deep Space 2" that were launched with the same rocket and accompanied the lander on its journey to the Martian surface, where they should have impacted in a distance of 60 km. A "Report on the Loss of the Mars Polar Lander and Deep Space 2 Missions" from March 2000 stated the most likely cause of the loss of the lander was a touchdown sensor in a landing leg that was spuriously activated, and based on this reading, the lander controls prematurely cut off the landing engine, assuming that it had already touched down, while it was still 40 m above ground. The impact of the lander in this case had a velocity of 22 m/sec instead of the 2.4 m/sec for which it was designed[*]. If this explanation is correct, Mars Polar Lander was also lost due to a communication problem between system components. One component caused vibrations or similar tremors that another component misinterpreted as a valid signal, which then lead to a chain of programmed responses. The impression is that the teams involved in the development should have talked more with each other to identify the risks and find solutions to avoid them and make the system failsafe.

---

[*] (Casani et al., 2000).

Solid teams are much easier to manage. Colocation, placing a project team at one place, is considered a good practice for project team success. Famous examples are the Lockheed Martin Advanced Development Center, the Skunk Works, near Los Angeles, USA, for military aircraft (since 1943), Porsche's Development Center in Weissach, near Stuttgart, Germany, for sports cars (since 1971), or the Gotenyama facilities in Japan from 1946, where Sony developed the first commercially successful transistor radio, launched in 1955, and laid the groundwork for the modern consumer electronics industry. Colocation shortens the way between team members to coordinate their work and build team spirit, eases communications, reduces bureaucratic overhead and the need for formal meetings, and simplifies the identification and resolution of misunderstandings.

There are also many examples of development teams distributed around the world. Their managers like to show off with the 24-hour development they do, as at any time someone in the world is working on the project. They use phone conferences and other technologies a lot, but the communications do not occur as often as in colocation. With time zone differences, someone has to get out of bed to attend. Face-to-face meetings when technological solutions prove insufficient require travelling, which is expensive, and the time needed is mostly unproductive for key staff. When companies see the need to cover costs, reducing travel is often among the first decisions, irrespective of the effects that this will have on their distributed project teams.

Culture is another issue. In a face-to-face setting, cultural issues are easier to avoid and to fix if someone puts "a foot in it". In a collocated environment, it is also easier to forge a team from members from different organizational units, or even different organizations, to enhance team spirit. Protection of classified data, by the way, is also easier in solid teams; employees of contractors, possibly in remote locations, can take away information much easier than people from the performing organization located together.

Is every virtual team a siloed team? Not necessarily. Sometimes the members of a project team have worked together in the past and are now dispersed across units, organizations, or locations. The strong rapport that these "old boy networks" (that may well include girls) have developed in the past may then be sufficient to keep up the solid spirit even over organizational or physical distances. Collocated teams in turn may be quite siloed: Staff members from different organizations or units have been brought together, but while the physical proximity is high, the organization is not, when the team members are instructed to follow different business objectives and have their own, often hidden, agendas.

Is colocation a "best practice"? Sometimes not. In some projects, proximity to local markets may be more important than among the team members. Facebook, for example, does most of its development work in Menlo Park, California, USA, but has worldwide teams distributed over 53 other locations to run projects together with their multinational clients. These offices are dedicated to customers, but also to lobbying and other forms of political influences that also require closeness to a target audience more than among the team members. Another reason for the preference of distributed teams may be the desire to tap hot spots of skills near to Universities or to companies that can make valuable staff redundant, or simply to places where certain resources are easier and less expensive to acquire.

Siloed teams should be managed with care. The cost benefit from distributed development has often turned into a monetary nightmare, when the costs of managing the teams exceeded the expected savings. The benefit from contracting work that an organization does not want to do with its own staff may vanish when the contractor causes new problems, for instance, because the company has financial problems. Technology may help to "solidify" teams, but the dangers are high of replacing human interaction with digital solutions.

The problem of team integration at NASA has been addressed in a second report by the Mars Climate Orbiter Mishap Investigation Board. While the first report had its focus on inadequate processes, the second focused on people. The report recommends to support (or replace) the "faster, cheaper, better" approach with a new one called "Mission Success First". It recommends to put the common interest of a project team over the particular interests of its team members[*]. In 2015, NASA had multiple success stories. Its Mars missions sent data and images from the planet with an impressive level of detail. NASA published images of the micro-planet Ceres and also of Pluto and its moon Charon that showed incredible detail, celestial bodies that before had only been blurred spots on images from observatories, and the scientific revenue was immense. With these successes, claims that money invested in their space programs would be wasted have become much rarer. Good project management is a great tool to ensure project support by stakeholders, and solidifying teams is a good way to achieve this goal.

The dynamics of success and failure dictate that we will often not have the time and budget, but possibly also not the strategic support of upper management to solidify the teams involved. Some organizations have developed solutions for such situations; the most radical was probably Volkswagen with their Modularer Querbaukasten (MQB), a modular kit for lateral engine placement, and Modularer Längsbaukasten (MLB), a modular kit for longitudinal engine placement design approaches. In essence, each is a collection of building blocks that have been developed independently and asynchronously. Volkswagen developers can take and swiftly re-assemble them to design a new car, in contrast to platform approaches that build one model based on modifications made to another one. The building block approach is used across its different brands (VW, Audi, Porsche, Seat, etc.) and has helped Volkswagen become the global market leader in passenger cars by numbers in 2014[†] and number 2 on the automotive world market by revenue[‡] after Toyota.

Volkswagen is also a good example of the risks that come with the approach: In September 2015, it became known that software used to control engine functions during testbed simulations was illegally altered to reduce the NO_x contents of Diesel motors in laboratory emissions testing, but reduced the effectiveness of the NO_x cleaning mechanisms during normal driving. The motor, an essential element of both the MQB and MLB design approaches, drives an estimated number of 11 million cars worldwide, built between 2007 and 2015. In combination with a result-driven management approach ("I do not care how you achieve the goal as long as you achieve it, and no, there is not more budget than what has already been allocated!"), a small group of engineers in highly siloed development projects was obviously able to secretly jeopardize the future of the corporation and its almost 600,000 employees by seemingly meeting emission goals, while no one seems to have asked how they met them. They placed a time bomb in the motor that went off in September 2015. For decades, Volkswagen staff repeated in conversations an aphorism, "If the company knew what the company knows", and the events of September 2015 showed how much truth the statement holds. When a modular system, which needs trust in the functioning of each element,

---

[*] (Stephenson. et al., 2000).
[†] (Statista, 2015a).
[‡] (Statista, 2015b).

collides with a high-pressure environment, which nurtures a culture of distrust, hidden agendas and finger pointing occur almost naturally, and concealed time bombs threaten the long-term success of project deliverables and their owners.

A positive observation at Volkswagen is that the company quickly identified the root cause of the problem and took immediate measures to change this culture. Whether they will succeed in changing a behavior that has spanned decades, with a sustainable effect, will be an interesting observation in the following years.

As project managers, our responsibility is limited to temporary endeavors, not to entire organizations that are expected to last for decades. The problems that we face in Siloed projects can be similar, however. Siloed projects can be successfully managed by carefully designing the interfaces between the different teams and their modules, but one has to take care that one unsound module may damage the entire system.

3.7 Blurred Projects and Focused Projects

There is a common conception of projects as endeavors with a defined moment of beginning and another as its end. They focus on achieving defined objectives, that are clarified and agreed upon, and management attention ensures that resources are available for the project as needed. There is, of course, a maximum number of resources that the performing organization is able to provide; nevertheless, project managers know which resources will be available to them and which will not. Start and end dates may need revision before their occurrence, and the project budget and the requirements that the project must meet may also be reconsidered and changed from time to time, but between these changes, the key data are well-defined and stable. In this type of project, the same clarity is assumed for the key players in these projects: There is no ambiguity about who is part of the project team, either as a worker, a decision maker in charge, or both, or each person's area of responsibility. In construction projects of this type, workers and other team members are allowed to enter the site; others are not. Developers and other team members in a focused software project get access to data, code, and design that others will not get. The people involved in the development of a new car model will know what it will look like, how fast it will run, how much space it will have, and so on; others in the company will not. The clear designation of people and things that belong to the project, of objectives that the project is dedicated to meet, and of the other key data about the project is an important concept in project management, and it is applied in many projects where this clarity supports the success of the project, if it is upheld to the end of the project.

There are other projects, which the experts described as "blurred". The reason for the blurring of the project may be ineffective project management, but the difficulties to separate the project from its environments and to produce clarity regarding objectives, people, start and end dates, etc. may also be inherent in the nature of the project. It may also be decided by senior managers. Clarity comes naturally with commitments, and organizational leaders often have difficulties committing to the projects they are performing in a reliable fashion and to the project managers who they in turn hold accountable for the results of these projects. Project managers then often have difficulties finding the balance between claiming too much clarity or not enough, between accepting the uncertainty they find themselves in or fighting a battle for clarity that the organization possibly cannot give them. Often, it is a problem in a weak matrix environment, where a functional organization uses its own resources for continuous production, services and other operational activities, and the demand for resources is hard to predict. If the market demands a large volume of the products or services, operational staff will have less time to work on the project. At other times, people may want to join the project team and influence its outcomes even though they were not originally part of the plan. This common unpredictability of demand can make it hard for organizational leaders to reliably assign functional resources to project work.

Many project teams have to multitask their work. From a manager's view, multitasking has the benefit of higher resource usage. For example, if a person is idle in one task because the task depends on yet undelivered input from another task, the person may be allocated to another task and stay productive. Multitasking also comes with the expectation that the organization can meet a variety of concurrent requirements with a limited number of highly flexible employees and contractors timely. In multitasking situations, one can often see resource assignments using percentages. Assigning generic resources early in a project by percentages can be helpful as a tool of progressive elaboration of a project plan. A statement such as: "We will need a developer for the project in March or April next year, and the person will have to dedicate 50% of their working time to the project" can be helpful when the actual assignment is still far away. Later in the project, when more information is available and progressive elaboration is applied to clarify the assignment, the generic resource would be replaced by a specific one, and the plan would be refined, as shown in Figure 3-7.

This process can be considered normal. Problems commonly occur when the percentage-based assignments are still used for the refined plan and during execution. When one asks what the 50% assignment means in reality, one rarely receives a clear answer. The agreement for the availability of the team member remains blurred. If a team member has been assigned for a 40 person-hour activity to work over four weeks (160 h) two hours per day, the project manager can expect that the person will be available for that time, and any changes in the availability would need to be agreed upon with the project manager. The project manager can also track the progress and identify early if the 40 work-hours estimate turns out to be sufficient or not. If a team member is assigned for 25%, no agreement has been made when the work will actually be done, and if the team member has not done anything for the project in the first two weeks,

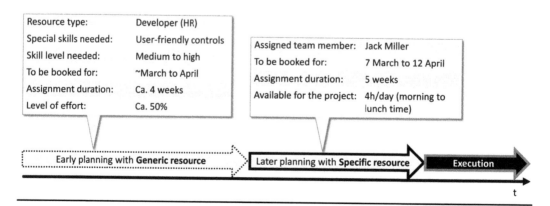

Figure 3-7 Progressive elaboration using a generic resource. This process can be considered.

there are still two weeks left for the activity, so the project manager cannot complain. We discussed in the second chapter the learning curve for the project manager and how important it is to know things early and to be able to make decisions, when there are still many options available at a relatively low cost. A common sign of a blurred project is that knowledge becomes available later, and decisions are therefore also made later, at times when project managers have to choose among a small number of options that have become expensive.

A company, known as Weasel Ltd. for this case story, taught me another interesting lesson some years ago on focused versus blurred team assignments. Often overlooked is the uneasiness for team members in unclear chains of commands when they have to work for the project (or several projects) and the functional organization at the same time. A team member complained about contradictory instructions from the project manager and from the person's functional manager. The problem was not caused by macro decisions such as business travel or major purchases the person had to make, but instead the cause was micro decisions that came several times per day—e.g., requests for urgent reports. The person complained that it was difficult to be a servant of two masters, but the blurred assignments made things worse, because the instructions from the two "masters" too often came at the same time and competed for execution.

The resolution was quite simple, implementing Figure 3-7: It was decided that the person worked for the project manager daily four hours before lunch and for the functional manager in the afternoon. The person knew at any moment for whom execution was expected, and the project manager could rely on the agreement and that the team member spent four hours of time working for the project*. The functional manager first rejected the agreement, probably fearing that it came with a loss of influence, but turned to a positive attitude after a while, when the benefit became obvious that the project would not disrupt the team member's work in the afternoon. It turned out, of course, that sometimes the agreement became too stringent when certain tasks demanded the team member's uninterrupted work for more than four hours a day. As the agreement was clear and simple, changing it was also easy, in most cases, it just took a phone call to make quick changes to the agreement. The most interesting observation was that un-blurring the assignment reduced stress for the team member.

At Weasel Ltd., it was also interesting to see that the common over-assignment of team members ended. Percentages are something abstract, and organizational leaders that allocate their human resources using percentages commonly over-allocate them by over 100% across functional tasks and projects. While this effect is not rare at all, companies often ignore the logical effect: phantom resources that are used in plans but will not be available for the projects. Phantom resources are a common cause for the inability of projects to adhere to schedules and meet their deadlines.

Percentage assignments of human and other resources are not always a poor practice. Early in the planning process, or later in the process in situations when information on resources availability is still uncertain and vague, they can be helpful. Eventually, during progressive elaboration of the plans, they should be replaced by more accurate and robust information to allow for planning and progress tracking and for predictions on the remaining work.

---

* Another observation that I made quite often was that a project manager believed that a team member, an organizational unit, or a contractor doing work for his or her project was unaware that the resource was busy with different work. The information on the unfinished work and the delays then came late, leading to the problems of late decision making already discussed.

Weasel Ltd. provided another example of how projects can get blurred. The company was a family run, mail-order company established many decades earlier and well established in its home market. Over the years, it optimized its business processes around a guiding principle that they called the "Catalogue Regime". The company traditionally published and distributed two thick catalogues a year, one for summer and the other for winter business, plus a number of small catalogues over the year with special offers to allow clearing stock at the end of a season and intermediate responses to competing offerings and also some smaller market-specific catalogues, such as for gardening items or sports articles. The main catalogue had a thickness of over 1,200 pages, was over 1.5 kg (3.9 lbs), and contained between 15,000 and 20,000 offerings. The catalogue regime was developed in adaptation to the cycles of the clothing industry with its summer and winter collections and dominated all key business processes in the organization. Supporting it was a dominant requirement for the software in use, which was a commercial off-the-shelf enterprise resource planning (ERP) solution, tailored with a considerable amount of individual customization with a lot of home-made software added to it. By the turn of the century, the whole company had become a massive and heavyweight player in its market, trusted by customers and shareholders and a well-known name for its buyers for all kinds of goods that one could buy.

The catalogue regime was the center of gravity of the company; almost the entire business of Weasel Ltd. was orbiting around it. Contracts with suppliers were made to ensure reliable delivery at fixed prices during the six months that a catalogue was valid. Product innovation also had a six-month rhythm, a successor product could only replace an older one when the new catalog was published; if a new product was launched on the market just after the editorial deadline of the catalogue, it would have taken Weasel Ltd. half a year to offer it to the market. This was not a problem in decades when the business world was less complex, and many manufacturers geared their product renewal cycles to the catalog dates. The company was ahead on another dimension since it had its warehouse and road logistics processes accelerated that it typically did not send customers acknowledgements for incoming orders, they were just dispatching the goods three to four days after order receipt; order confirmation letters would not have been faster*. Mail order by the turn of the century was one of the industries that was successfully built on a strong foundation of automation and optimization.

Then came Amazon.com. It started in 1995, as a small online bookstore during a time of Internet hype that ended in the burst of the dot.com bubble in the year 2000. Amazon's business model allowed them not only to survive the burst, but also to diversify into a full-fledged digital warehouse and provider of digital services. Similar to Amazon, a large number of small and medium online shops grew as well, and some of them began selling their products on Amazon.com as an online platform, while others did it independently. Together, these companies changed the rules of retail business and accelerated the pace of change and innovation dramatically, and the mail-order business became true e-commerce.

The entire vending process from product selection through tracking the delivery and payment is now done online, the only activity not done online is the physical dispatch process. The speed of the new business outpaced Weasel Ltd. and similar companies, who seemed prompt and speedy before that time but now created the impression of being rather contemplative and inert. Online customers order over the web and expect dispatch on the next

---

* A student of mine from the company once told me in a class that the company was actually faster delivering goods than sending letters.

working day, sometimes even on the same day. The online business is much more driven by frequently changing prices, sometimes several times a day. For many products, especially electronics, prices have a tendency to fall over months until a better successor product is launched at a higher price, which is then also bound for decline over the next months. For many products, a distinction is made today between a "catalogue price" and a "street price", which is not the price on the street but on the Internet. Also the information available to the customer has increased dramatically. Buying with a catalogue saved the customer from having to spend time and money walking through shops to compare prices, but on the Internet, price comparisons have become easy and simple, and of course, there are special websites that further speed up the process. Customers also rate products to give other customers hints on the perceived quality and usefulness of products, something a paper catalogue could never do. The catalogue regime with its six-month freeze cannot compete with the small intermediate catalogue,

Another ace up Amazon's sleeve is its partnership program with independent, mostly small vendors who can sell their goods using Amazon's infrastructure, often directly competing with Amazon's offerings. Amazon offers them its services for logistics, billing, and payment. It seems counterintuitive at first that Amazon supports competing sellers, but the benefits they gain are worth the cost. Amazon earns a percent of the sales of the partners, and also gets information on how new products are selling and at what price. Amazon adds products to its own product portfolio that are already successful, and rejects those that are not. Amazon also gets information on its customers' purchases from other vendors and can target its advertisements to specific customer requirements. The only way a traditional catalogue company can learn how well a new product is selling is by adding it to the catalogue. Sometimes, Weasel Ltd. rejected a product that became a bestseller with other vendors; at other times, it filled its warehouses with shelf warmers, items that no one wanted to buy. Weasel Ltd. could not compete with Amazon's speed or with the information on products and consumers that Amazon had. Peter F. Drucker once wrote that if a company has a 30% market share, it is a giant in its market; however, this means that it has no knowledge of the other 70%, the share of those who are not the company's customers[*]. A point not addressed by Drucker was that these other 70% have no knowledge of what the company can do for them. Through their partnership program, Amazon gets information on these 70% as well, and these buyers also know Amazon. From time to time, Amazon will lose business to sellers in their partnership program, but as the customers do business inside Amazon's bailiwick, the customers are not lost for the company in such a moment. Being a closed warehouse and an open marketplace at the same time is a highly powerful business model[†].

So why not simply drop the catalogue regime and move the business to an online platform as well? Weasel Ltd. had too many legacies to easily do so. First, this approach would have meant risking the traditional customer base. Not all clients were happy to order over the Internet, many still loved the catalogue and still do. Among them were many older and disabled people, who had difficulties leaving home and were happy to have an alternative. Many of them would not want to use Internet services. While this group of customers declined over the years, they were still important for the mail-order company's business. These people were much less interested in daily price changes and in the rapid delivery that e-commerce

---

[*] (Drucker, 2013, p. 84).
[†] It is no surprise that some people consider Amazon's power as too strong.

promised, but they would also not have wanted to buy a product from the catalogue at twice the price for which it was meanwhile sold online. Customers in such a situation could also have taken Weasel Ltd. to court for discrimination; an allegation that the company would want to avoid by all means. It preferred to stand for the vendor who made customers happy, not angry.

Another obstacle for a quick change were the internal processes of the company and the software that supported them. They were highly optimized and had a lot of customization. IT experts know the basic problem with customization: it makes it difficult to make changes to systems. The vendor of standard software can offer customary routines that manage upgrades and changes to the standard components of this software, but for customized components, the owners of the software must manage changes by themselves. In order to change the systems, Weasel Ltd. had to rely on the availability of the people who originally developed the customization, often years ago, on their ability to remember what they actually did when they developed the customization, and on the completeness and correctness of documentation, which is rarely a priority in customization projects with tough deadlines and budgets.

One should also not underestimate how much these processes influenced the culture of the organization that was not open for a revolution. To implement change, Weasel Ltd. needed younger staff members, who often were not welcomed by existing employees. The corporation had started an online presence in 1995, not as an e-commerce platform but for promotion and to win new subscribers to its catalogue. The website was a fringe activity that did not touch the core business at all. But developing a new core business without damaging the old one was difficult. One of its major competitors tried to avoid the change, adhere with the old business model, and went bankrupt in 2009. Another competitor, much more enthusiastic about the opportunities that came from online business, changed its core business rapidly; too rapidly perhaps, moving all market activities to the Internet in only a few years' time, and then went bankrupt as well.

Weasel Ltd. had some projects launched to carefully add online business to its traditional sales model. These projects could not simply focus on their own deliverables and be on time, budget, and scope and "put mission success first", but got blurred by the need to continuously balance their contents and their pace to the changes that happened on the market and inside the organization. Weasel could learn some lessons from Amazon.com, such as developing a storefront for partner companies, but also needed to take care of its legacy business. In addition to operational effectiveness it had to develop business agility and adaptiveness to a fast-changing market with contradicting demands. The company hired external consultants and developers, who first had to develop a deep understanding in the way Weasel ran its business with the purpose to find new solutions without impacting existing ones. It seems Weasel Ltd. was finally successful in managing the split business. In 2013, roughly two-thirds of the business was done using the online channel, and the trend is clear that its future is in the online business, as a competitor of Amazon, which is business-wise more than 10 times larger and with faster growth, much less restricted by consideration for legacies.

A blurred project may not necessarily be a bad thing. It may also be a project that is mindful about the consequences of its performance and its deliverables and that coordinates its scope and pace with the organization for which it is performed to avoid damages to the organization and ensure business benefits.

3.8 High-Impact Projects and Low-Impact Projects

In the introduction to this chapter, I discussed the project called Heathrow T5, the terminal building for British Airways at London Heathrow, which opened in March 2008, and how its operation collapsed immediately after opening, and with it, practically the operation of British Airways. As a passenger of British Airways flight on the next day, I could see the disaster there with my own eyes, which was extremely unpleasant from the viewpoint of a traveler, but highly interesting from that of an observer interested in the subject matter of troubled projects.

Many project managers rely on their many years of experience, which are of course an important element of their professionalism, but two other important elements are education and observations. The interesting thing about observing other projects is that one does not have to make all mistakes by oneself. When others learn a lesson the hard way, one can learn the lesson as well without having to make the same mistake. Heathrow T5 is a definite example. What lessons can we learn from the case?

There were many analyses written on this major disaster; most of them focused on technical issues. A major problem area was obviously the luggage transportation systems that did not function as planned. The glitches were on hardware and software side, and the breakdown resulted in 42,000 pieces of baggage that weren't returned to their owners until weeks later. It took British Airways two weeks to get back to a normal schedule, during which it had to cancel over 500 flights. The financial damage was enormous. It had to pay back fees to holders of tickets on canceled flights and cover the extra costs of tickets bought by passengers on other airlines in the short term, which were very expensive (like mine). It lost business from travelers whose trust in the airline was destroyed, and it had to cover the costs necessary to rework and repair the faulty systems. Still months later, it was reported that about 8% of the bags of travelers changing flights at T5 were lost.

There was a multitude of technical glitches and human errors that in combination caused the disaster, but there seems to be a main story line of events where the most calamitous mistakes came together.

The story begins during the planning and the construction of T5, which was actually two projects: First, a construction and infrastructure development project performed by British Airport Authorities (BAA), which, despite its name, was not a government agency but a privately held company as of 1986. In 2012, the company's name was changed to "Heathrow Airport Holding", and it functioned as a landlord at the airport.

The other project was performed by the tenant, British Airways (BA), also a privately-held company, and had the objective to enable the services necessary to operate the terminal. This project was mainly concerned with recruiting, organizing, and training people and arranging their working environments. The two projects were tightly linked, and it may have been helpful to put a program manager on top of both projects to integrate them.

This approach could have prevented the fragmentation of the joint program and competitive behavior between the projects, but such a position was obviously not implemented. Establishing a position on top would have meant BAA and BA giving up a degree of independence and sovereignty. The egos of executives at both companies may have been too big to follow such a program approach.

Measured in classical metrics of scope, time, and cost*, the construction project was successful, delivering a large airport terminal on time and budget (£4.3 bln., U.S. $8.7 bln. in 2008). The formal risk management activities applied before the hand-over were enormous, and they were a popular subject for articles in special-interest magazines and for speakers in project management association events, but some risks were overlooked. One was the risk of game-theoretical dilemmas that are typical for situations where two organizationally independent, but subject-wise related projects are performed at the same location and call for conflicts on the technical, organizational, and inter-personal levels. Another group of risks was ignored because they lay beyond the event horizons of quality and risk management of the project managers.

Event horizons for quality and risk management mainly stem from the fact that one cannot consider all possible factors of uncertainty. The consideration of uncertainties is often restricted to those that are known from experience or mandatory by law or contract. Some people are better in planning ahead and considering a large number of uncertainty factors than others, but all humans have limitations in this competency.

Quality and risk management are mainly constrained by two elements: lack of time and budget, and domain restrictions. As an example of the first, testing and inspections are habitually regarded as disposable reserves in many projects, especially in those that are running out of time and budget. Both come late in most project workflows and are expensive as the time and money left may not be sufficient to do them as necessary. To make things worse, both may lead to rework that will cost additional time and money. So, they are often reduced or completely abandoned when deadlines and monetary limitations restrain the project. Domain restrictions refer to local or organizational distance, when certain activities that are relevant for the project are made by another organization and at a location that is too far away or hidden behind walls.

Lack of quality and risk management is easily discernable, when managers act based on trust in the immediate functioning of a complex system. Has it not been thoroughly designed and built by expensive professionals? This trust then replaces time to prepare for errors in the system. Instead of an attitude that assumes that the deliverables most likely have faults that must be found and fixed, with the possibility of a positive surprise when things work correctly from the start, a "Frog King"† attitude is then taken that things have to work correctly from the start, of course, with the prospect of painful and embarrassing surprises, when they do not.

At the Heathrow T5 project, the parking zones for the staff were a technical root cause of the problems in March 2008‡. The construction of the parking areas was a separate project and while there was a lot of testing done during the project for the terminal operations, the parking space was obviously neglected. According to reports, there was a lot of confusion in the parking lots on the morning of March 27. Drivers in search of free spaces blocked others, who had the same intention, which resulted in a massive traffic jam inside and outside of the lot. Once a staff member had parked the car, the person had to pass the security checkpoints, where other staff members meanwhile stood in long waiting lines. The security checks were also slow, and,

---

* Sometimes referred to as "triple constraints", "magic triangle", or "iron triangle". Different sources put different aspects into the corners. T5 is an excellent example of the fundamental insufficiency of the model.

† The first sentence in Grimm's fairytale *The Frog King* says: "In olden times, when wishing still helped one . . .".

‡ A brief description of the situation was published shortly after the initial event (BBC News, 2008).

to make things even worse, many security staff members were still outside the building in the traffic jam, trying to find a parking place. One can imagine the conflict between BA's luggage workers who knew that BAA would soon start the belts and the understaffed security adhering to their rules and trying to keep things under control at their checkpoints. And then, the belts started and brought an avalanche of luggage to a small number of persons. The luggage system did not work perfectly on its first day of operations, so some free staff would have been needed to work around and fix the errors manually, but there was not even enough staff there to handle normal operations.

What would have been a better approach to making the start of operations safer at Heathrow T5? Of course, it is always easy to make recommendations from hindsight, but in production and service industries, a handover of a readily commissioned complex system is normally followed by a ramp-up phase of several months. A ramp-up phase is a period beginning with low productivity, which is carefully increased over time until it reaches its intended maximum. Ramp-up phases have two benefits since they avoid a situation where the organization is flooded with many erroneous parts, and they leave some resources free to fix errors or exploit opportunities for improvement that become visible during early operation. Figure 3-8 shows how a ramp-up phase is an overlapping of the project life cycle with the operations life cycle.

In Heathrow T5, it may have been helpful to start with a relatively small number of flights on the first day, leaving capacity free to deal with any possible technical problems as they occur, and then carefully ramp-up the number of flights over the next few weeks, until all flights had been relocated from the old terminals to the brand new one. Such a solution had been chosen in 2006 in Bangkok for the new Suvarnabhumi Airport, which began its operations with only a small number of flights on September 15, 2006. Problems materialized with the baggage system that could soon be fixed, and over the next two weeks, other airlines moved from Bangkok's old Dun Mueang Airport one after the other. On September 28, the airport seemed to be working effectively and was officially opened to run at full capacity from that time. In the next weeks, a number of new problems emerged, so a decision was made in February 2007 to reopen the old Don Mueang Airport for domestic flights. A ramp-up phase for an airport

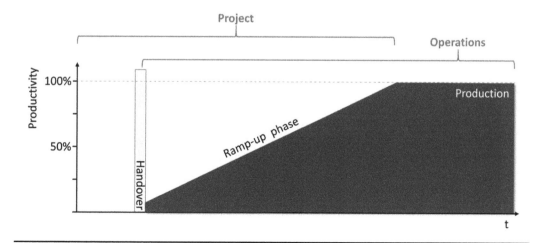

Figure 3-8   A ramp-up phase reduces the risks in new production environments.

includes keeping an old facility functional as a fallback solution in case of problems and being careful and watchful with the new system to identify how much load it can bear. With the reduced number of flights, Suvarnabhumi Airport seems to be working well today.

Why did the managers responsible at Heathrow decide to not choose the option to slowly ramp up the traffic in Terminal 5 by moving flights in small portions from their old Terminal 1 to the new terminal? Instead, they even planned to relocate a number of flights from Gatwick Airport, their other hub near London, to Heathrow T5 only three days after the terminal opened, and as a last step to move almost all long-haul flights to T5 by end of April the same year. The answer seems to come down to three things: necessities, cost, and egos. The necessities have their origin in the function of Heathrow as the central hub for British Airways, which runs a business model to bring travelers with smaller, short-distance aircraft to Heathrow and have them change there to the large intercontinental flights and vice versa. Having operations distributed over various terminals, one of them, in Gatwick, 58 km (36 miles) away, is more difficult, causes delays and dissatisfaction for travelers changing flights and increases costs for the transportation of these people and their luggage. The desire of BA management was strong to collocate the traffic in just one terminal. But there is another reason: During the night of May 15 to 16 in 1992, the entire Münich Airport was moved 40 km (24 miles). The transition of the conclusion of operations at the old location and the start of operations early next morning at the new location went smoothly. Hong Kong also moved an entire airport 30 km (19 miles) overnight from July 5 to 6, 1998. Concurrent operations of both airports for ramp up and as a fallback solution did not seem feasible at these places, as expensive airport inventory would have been needed twice at the old and the new location, and tight flight schedules of travelers with connecting flights departing from the other airport would have failed. In Münich and Hong Kong, moving overnight and starting operations at the new airport was just a bare necessity.

This necessity did not exist in Heathrow, where all terminals shared the same airport infrastructure. The infrastructure of Terminal 5 was new, including the baggage handling system; the old terminal remained in operation until June 2015, so its infrastructure had to remain in place. The distance between the terminals is only 2.7 km (1.7 miles) and takes just nine minutes by Tube train (London Underground), which runs every 10 minutes. An additional temporary shuttle service for travelers and their luggage would have been another feasible option to overcome the difficult first weeks of operations of T5, and as much as BA had disliked the old terminal, it would have been a perfect fallback solution when problems occurred. Managers at Heathrow obviously made a decision that if Münich can move an entire airport overnight and restart at the new location without a ramp-up phase, BA and BAA would also be able to do this with "only" a terminal. It seems, egos stood in the way of making such a decision, probably the true root cause of the problem. Egos can be especially difficult in chicken race situations: One of the two projects (or their project managers) at BA and BAA should have made the proposal to start the terminal operations slowly and ramp it up watchfully. I spoke with some people involved in the project, and it seems plausible to me that the managers of both projects saw themselves racing to the edge, but no one wanted to be the chicken and jump out of the car by talking about the concerns. Still, today, when one talks with BAA managers, the root cause of the disaster, in their opinion, was the unprepared and undertrained BA staff. BA's people instead put the blame on the faulty infrastructure and point at the resignation of BAA's Airport Chief less than seven weeks after the event as a confirmation of their position.

Terminal 5 had huge impact on the businesses of both BAA and BA. Its success, when the crisis was finally mastered, helped to streamline and improve the businesses of both companies, but the weeks of crisis in 2008 also had a dramatic impact on both organizations. The United Kingdom felt ashamed as well; British Airways is its "flag carrier", the airline that is allowed to present the Union Jack on the fins of its aircraft, and Heathrow is Britain's largest and busiest airport. High-impact projects require more care when basic decisions are made, and risk management does not only require certain formalisms, but also watchfulness, and contingency and fallback plans in place when expected or unexpected risks occur. High impact means high stakes.

3.9 Customer Projects and Internal Projects

I discussed the difference between customer projects and internal projects in Chapter 2. A field expert observed that a practice had worked well in one project but failed in another one, related to this distinction. It seems the most obvious, even most trivial distinction in project management; nevertheless, the differences between the two types of projects are often overlooked or, if they are mentioned, they are not elaborated with all their fundamental details. Customer projects are mostly performed as profit centers under contract with a customer, which are two influence factors that move them toward a strong matrix organization. Internal projects, in contrast, are more likely to be performed in a weak matrix, as they are cost centers that do not directly create income (their deliverables may do so later) and as they compete for resources with the profit centers inside the performing organizations, which are in most cases the functional business units that carry out the money-making operations.

I wondered about the percent of project managers assigned to customer projects versus internal projects and again designed a small survey for project professionals. Over 11 days, from August 30 to September 10, 2015, I received 246 responses that were distributed as shown in Figure 3-9.

The survey showed that the assignments of project managers are almost evenly distributed over the two types, at least among the respondents of the survey, which generally confirms the observation that I make in my classes. The option for other project setups allowed comments by survey participants, who responded with statements like "both internal and external" and "I'm kind of a sponsor for an internal project as well as responsible for several customer projects".

Another difference in managing a customer project depends greatly on the contract type. To make things more complicated, there are two different kinds of legal definitions of contract types and understanding both helps better manage the customer project. There are different definitions used in Common law countries, where the Anglo-American legal system applies, and in Civil law countries, including Continental Europe, most parts of Latin America, Africa, and parts of Asia. In Common law, the distinction is made based on the obligation of the customer to pay:

- Fixed price contract
- Cost reimbursable and time & material (T&M) contract

Both types can come with or without conditional bonuses (incentives) or price deductions in form of liquidated damages (LDs) or penalties.

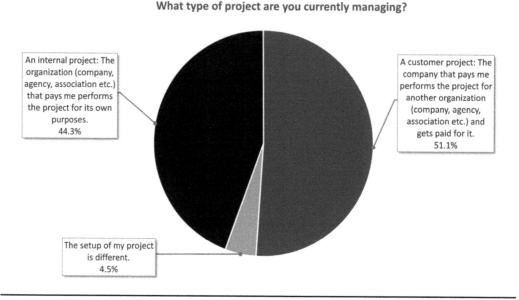

What type of project are you currently managing?

An internal project: The organization (company, agency, association etc.) that pays me performs the project for its own purposes.
44.3%

A customer project: The company that pays me performs the project for another organization (company, agency, association etc.) and gets paid for it.
51.1%

The setup of my project is different.
4.5%

Figure 3-9   Distribution of assignments of project managers in internal projects vs. customer projects.

In Civil law environments, contract types are based on the obligation of the vendor:

- Result contract (French: *Contrat d'entreprise,* German: *Werkvertrag*)
- Service contract (French: *Contrat de moyen,* German: *Dienstvertrag*)

Bonuses are much rarer in Civil law countries, but price deductions in the form of contractual penalties are common.

A result may be defined as a tangible product or an intangible deliverable. It is specified in a result contract that the contractor must supply the deliverables to have a breach of contract situation. In a service contract, the contractor must provide certain resources that help the customer to create the result or attempt to create the result for the customer, but it is not in breach of contract if this attempt does not succeed. There is a common opinion that fixed price and result contracts are the same, and also service and cost-reimbursable/T&M are identical, but the contract definitions can be mixed as shown in Figure 3-10. Software licenses, for example, are commonly bought for a fixed price but do not give the customer the ownership over the software, just the privilege to use it[*].

The lowest risk for the contractor is in a cost-reimbursable contract for a service. The contractor must ensure the availability of the resources that are necessary to provide the services, and the customer pays the costs of these resources, plus a type of fee. Incentive fees may add some risk to the contract, but at least the resources are paid by the customer. T&M contracts have higher risk for contractor since the customer pays defined rates for the resources that may not actually cover the rates that the contractor has to pay for them. Risks increase with the

---

[*] I remember the time when phone contracts were paid by minutes that the telephone line was used by the customer multiplied by a certain rate per minute. By that time, a service contract was also a T&M contract.

fixed price that was defined based on early assumptions that may not be true when the project is being performed. On the other risk dimension, jeopardy goes up for the contractor when the obligation is to deliver the result, not only to provide the resources.

Project managers in internal projects do not have to be concerned about these questions, except when they procure project work or work results from outside the organization, in which they sit on the other side of the negotiation table. As contractors also procure work to subcontractors, this is then a common area of interest. But as internal projects are not run under contract, contractual questions seen from a vendor side are not an issue that an internal project manager has to consider.

The contractual aspect of a customer has generally another corollary since customer projects are harder to terminate than internal projects. In internal projects, terminating a project is a business decision, and after the decision has been made, one may see some people depressed and demoralized, while others celebrate. Customer projects are performed under contract, typically with written contracts in most cases that protect the interests of both parties. For instance, the customer expects a set of deliverables available for use after some time, and the contractor has blocked resources for the project and ordered items, often long term in advance, and the subcontractor or vendor also wants to be paid. The contract is therefore negotiated for most projects in a way to bind both parties to meet their obligations[*]. In some jurisdictions, termination clauses "for convenience" are common practice. I know them mainly from public projects in the U.S.; they allow a customer to terminate the contract even in the absence of fault or breach by the contractor and without giving a reason. At times, the contractor can request a termination for convenience. It is unusual to have such clauses for the contractor, and they can be invalidated by a court if it finds that the termination was done by the customer or the contractor in bad faith.

I observed an example of the binding nature of a contract in 2012, when the training company Vole, Inc.[†] put in a bid for a major industrial customer, Shrew Corp., to perform a staff development project. The project included methodology training for several hundred project

---

[*] In larger projects, mainly from infrastructure and construction, issuing insurances, so-called performance bonds, is also common practice that can make it even more difficult for the contractor to step out of the contract, because the insurance carrier will claim its payment to the customer for the nonperformance from the contractor.

[†] The actual names of the companies cannot be disclosed.

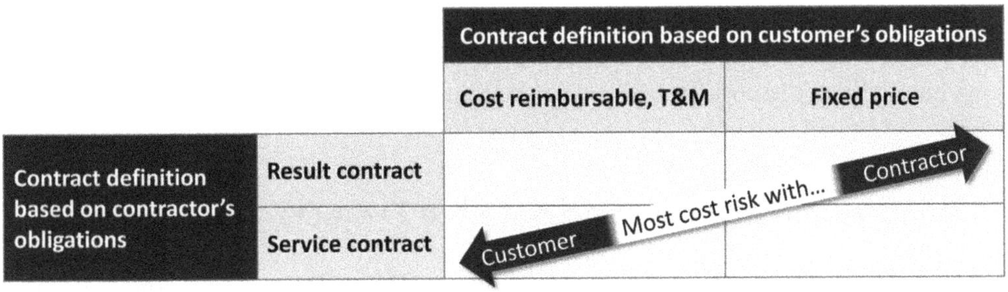

| | | Contract definition based on customer's obligations | |
		Cost reimbursable, T&M	Fixed price
Contract definition based on contractor's obligations	Result contract		Contractor
	Service contract	Customer Most cost risk with...	

Figure 3-10   Risk allocation is based on two things: obligations of the customer and obligations of the contractor.

managers and members of project management teams over a period of two years and included classes on two subjects, the proprietary "Shrew" methodology and Scrum, an agile methodology discussed briefly later in this book. As I knew both companies and the proposal management as a discipline quite well, I was asked to support the proposal manager of Vole Inc. in developing a capture strategy as a consultant. The proposal process went well, and Vole finally got the contract against aggressive competition on a T&M basis.

One problem that I saw during the bidding processes was that the rates Vole offered to Shrew Corp. were ones that I considered too low. Vole Inc. assumed that it would perform the project with its own staff and could run the project as a small-margin business without making a loss but as a way to open the door to Shrew, providing the opportunity to generate future business. I considered this "low-balling" approach dangerous. Shrew's requirements for the trainers were quite challenging, especially regarding formal education and the command of foreign languages on the level needed for seminars, and the risk that Vole's staff would not be accepted was quite high. My recommendation was to offer higher rates and, if the customer would not accept them, consider it to be a business that one should happily give to competitors to keep them busy and focus instead on developing more attractive business with other customers. This was not the decision that the company made; Vole offered at a low price. This is an observation that I have made repeatedly in customer project business during business development with new customers. Once a lot of time, effort, and energy has been put into winning the contract, fear increases of losing the competition and then having to explain to management why one has invested so much in the customer.

Vole Inc. finally won the contract, and the proposal manager became the project manager, because the company had no other person free for the job. My warning came true, the customer did not accept the employed trainers, and Vole had to find self-employed trainers on the market to replace them. These trainers' daily rates were about 50% over the rates that Shrew Corp. paid to Vole Ltd. After about nine months, Vole Ltd. found out that no follow-up business was to be expected. It then tried to get released from the contract, but it took another nine months to convince the customer to terminate the contract. I talked with customer representatives after this time, and Vole's reputation was then not positive for any future business. If a contractor needs to perform a project for a customer with too scarce resources, the customer will at the end of the project not remember the great service received for an inexpensive price, but it will remember a company that rushed from one problem to the next.

The lesson from this example: The project manager in a customer project is also a manager of a business relationship that the parties entered for mutual benefits. The tasks of paying attention to these benefits and balancing the often competing expectations and requirements of both parties is another difficult responsibility in this position. It requires specific practices and separates it from the position of a project manager in an internal project.

3.10 Stand-Alone Projects and Satellite Projects

Virtually all literature on project management assumes that a project stands alone. Of course, it has some dependencies to the project environment, which may come in the form of a matrix organization, an overlay of a more static functional hierarchy built for operations, and the dynamic team structures that we develop to achieve our project goals. Dependencies among projects may also occur when using a program structure. Because the project is part of the program, program deliverables and benefits must be created together with the other projects in the program. The project may also be performed inside a portfolio of projects, which may have independent goals, but share the same resources, including people, space, tools, materials, and the most important of all resources, management attention.

Projects may be linked in a different way. When NASA, for example, lost the already discussed Mars Climate Orbiter probe in September 1999, it also meant failure for two scientific projects:

- Mars Color Imager (MARCI) was planned to use two cameras mounted under the probe to observe the Martian atmospheric processes, study the interaction of the atmosphere with the surface, and examine surface features that conserved information on the Martian climate in the past.

- Pressure Modulated Infrared Radiometer (PMIRR) used an instrument to measure the thermal structure of the atmosphere, its dust and vapor loading, and more aspects of the Martian atmosphere to help understand its dynamics.

The Mars Polar Lander that was flying concurrently with the Mars Climate Orbiter and got lost during landing had five scientific projects on board, and the two Deep Space 2 probes that flew with it, and also got lost, had another four. Behind each of these 11 projects was a team that had spent years preparing and had expectations on the results that they could present to the global scientific community.

MARCI got a second chance with the Mars Reconnaissance Orbiter probe, which started in August 2005, and successfully went into orbit seven months later. MARCI has since sent data to Earth. This probe carried also a successor instrument for PMIRR, called Mars Climate Sounder (MCS). For PMIRR, the loss of Mars Climate Orbiter was already the second failure; the first was the Mars Observer probe that carried it also and had been lost in August 1993. It took the project team involved in MARCI two attempts, the PMIRR/MCS team even three attempts, the latter over 14 years, to bring their instruments into operations in orbit around Mars. The delays had nothing to do with problems in their own projects but with failures in projects with dependencies. One of the field experts in the research suggested calling these dependent projects "satellite projects", and later thought to call the projects they depend upon "principal projects".

I have not seen the concept of Satellite projects discussed in literature, neither in scientific literature nor in text books, but in cinema. *Once Upon a Time in the West* was an Italian "Spaghetti Western" from 1968, directed by Sergio Leone, and starring Charles Bronson, Henry Fonda, and Claudia Cardinale. It told the fictitious story of its protagonists dealing with two projects in the 19th century American Wild West near an also fictitious town called Flagstone. The first project was building a railway line coast to coast, and it was the principal project; the second, depending on the first, was building a railway station in an empty place called Sweetwater, 50 miles west of Flagstone, where the engineers could fill the water tanks of the steam engines, and where the station owner could make a fortune developing a city from nothing that derives income from the passengers of the trains that stopped in the city. The owner of Sweetwater bought the place years before and was waiting for the track construction to arrive at his location when the story begins.

Satellite projects like this one are not rare at all. I have already mentioned projects performed by corporations listed in U.S. stock exchanges in 2002 to 2005 to become compliant with the Sarbanes-Oxley Act of 2002 (SOX), which mandated that these corporations report data to shareholders—data that could affect the financial and business fitness of the corporation and the value of its shares. Under the law, these communications have to be true, the data have to be audited, and it has to be communicated quickly. Corporations spent billions of dollars on their SOX compliance projects. The projects were not discretionary—managers were threatened with being put in jail if the organization did not comply. The SOX compliance projects touched internal processes and software that had not been changed over years and brought workarounds back to mind that had been developed long ago as ad hoc fixes to cope with problems in these processes and software. Often, the process gaps were not a problem of the solutions in place but were inadequacies of the interfaces between them. As all the ad hoc workarounds somehow worked, they never were fixed in a controlled manner. These workarounds were often costly and time-consuming, but they worked somehow and allowed the corporations to shift resources to other things. Automation without optimization often leads to inefficiencies, which are hard to eliminate once they are in place. The SOX compliance projects forced organizations to have a closer look at these inefficiencies, and many corporations then ran satellite projects that used the insight from the SOX compliance projects to overhaul businesses and software and establish more efficient services. The SOX compliance projects as the principal projects had no business case, they were mandated by U.S. law, but the satellite projects had the business case to improve operational efficiency. Satellite projects may be able to add benefit to a principal project.

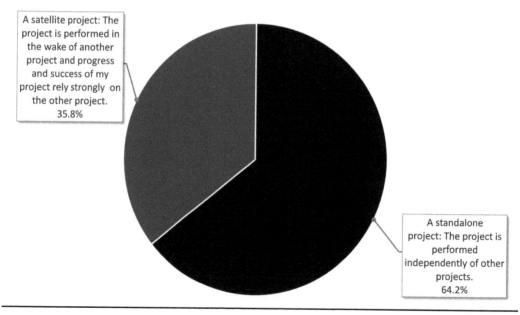

Figure 3-11 Distribution of assignments of project managers in internal projects vs. customer projects.

Similar effects could be observed in Europe with Single Euro Payments Area (SEPA), which forced businesses to modify their business software to cope with a system with unified bank account numbers across the entire Euro zone, very long ones, by the way, with 22 digits. Companies were not free in their decision if they wanted to change their systems to the terms of SEPA, it was required by law. In addition, not being able to pay or get paid would have been a prospect no company would accept, so they had to do the mandatory SEPA projects, but in their wake, many other system improvements were made, often in separate satellite projects.

Again, I wondered how frequent satellite projects are and asked practitioners in a survey whether they were managing a satellite or a standalone project. The results are shown in Figure 3-11. Satellite projects are less frequent than standalone projects but are not rare. When managing a satellite project, one must always have an eye on the principal project, because the success of the satellite project relies to a certain degree on it. Troubles, failures, and changes in the principal project must not remain unnoticed because they can have a major impact on the satellite. Risk management must also include consideration of the things that can go wrong in the principal project in addition to those on the satellite.

3.11 Predictable Projects and Exploratory Projects

In June/July 2012, I started discovering the diverse, situational nature of project management. The starting point was some deeply felt uneasiness on certain reports on project success and failure that postulated a measurable success criterion: "Project met its original requirements regarding budget, deadlines, functions, etc."* The assertion was that a project that meets its original requirements is successful; a project that does not is regarded as a failure[†]. Based on this metric, surveys were published that communicated incredibly high percentages of failed projects. My simple question was: What if the requirements have been changed during the life cycle of the project? When a project met the original requirements, should it not be seen as a failure then, because it missed adapting to the later ones that would be valid at the end? And a project that met the requirements that were valid at its end, would it not be regarded as a failure following the logic of these reports? In my past assignments as a practicing project manager, before I became a trainer, I had several projects that had to be performed against changing requirements, and the projects met these in the end. Being asked about one of these projects in such a survey, I would answer the question, "Did the project meet its original requirements?" with "No", and the reports would consider it a failure, which it obviously was not.

A second question was: What if the project requirements cannot be specified at the beginning of the project? An internal requester or a paying customer may not have enough special knowledge to formulate the requirements; the internal team or the contractor, on the other hand, may not have enough knowledge of the requester's or the customer's needs to formulate the requirements for them. In such a project where requirements need to be discovered while

---

* A popular example is the "CHAOS Report" published by the Standish Group (Standish Group, 1995, p. 2), but also PMI® (Project Management Institute) uses this metric in its "Pulse of the Profession" analyses (PMI®, 2014).

[†] The reports often use different terminology, but this is their essential claim.

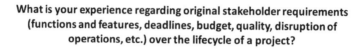

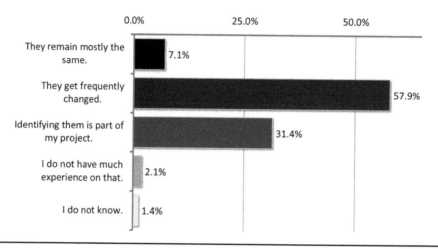

What is your experience regarding original stakeholder requirements (functions and features, deadlines, budget, quality, disruption of operations, etc.) over the lifecycle of a project?

- They remain mostly the same. — 7.1%
- They get frequently changed. — 57.9%
- Identifying them is part of my project. — 31.4%
- I do not have much experience on that. — 2.1%
- I do not know. — 1.4%

Figure 3-12   Responses in a small ad hoc survey in LinkedIn, June/July 2012.

the project is performed, how would one apply this metric? These projects do not have original requirements at all.

I wondered how often these situations occur. Could it be that my criticism related only to a negligible minority of projects? I decided to make a little ad hoc survey on the question, using a function in the social network LinkedIn, which was available by that time (June 2012) but has been unfortunately discontinued. I saved the data before they were made inaccessible, and the interesting results are show in Figure 3-12.

One should point out that this survey, as some more that I am using here, would not meet strict scientific requirements. It is rather a momentary illustration of the responses of a limited and self-selected group of people. The project managers involved may not truly represent all project managers and project situations. On the other hand, the participants came from the project management discipline, and the responses confirm my everyday observations as a trainer and volunteer in a global association, who is in professional contact with many project managers. It may not be a scientific approach, but it is an approach by a trainer who asks people when he wants to know something about them.

Projects with static requirements came out in the survey as a comparatively small group. The survey had 140 respondents, and over 50% selected that they had to deal with changes relating to the requirements during the project. The second largest group, almost a third, responded that they have to identify the requirements by themselves. A small group said that in their projects, requirements are mostly static. The small percentages for the answers, "I do not have much experience" and "I do not know" can be taken as a signal that the respondents of the survey came from the discipline and had enough knowledge and experience to make a valid selection.

For almost 90% of the respondents, "Meeting original goals" would not have been a valid success metric, as the requirements, which include these goals, were either changing during the project or did not even exist at the onset of the project. Success and failure seem simple when one takes a superficial glance at them, but once one starts digging deeper, one will notice that they are rather difficult to understand and are subject to complex dynamics.

The topic occurred in my research once in the form that a set of practices that was appropriate for a predictable project was detrimental for another one, which was much less predictable. We decided to call this other type of project "exploratory". In my research, the relevance of the topic was probably not fully shown, but I consider it highly important. We are again forced to "navigate between two monsters": predicable and exploratory projects.

3.11.1 Predictable Projects

Seven percent of the respondents in my survey said that this is the type of project that they are mostly familiar with.

These projects allow forecasts and predictions that reach far into the future, possibly until the end of the project. The project requirements are defined early and are comparatively fixed and inert. The project environment is rather static, availability of resources is reliable, and the other factors that influence the requirements rarely change. This inertness of the project and its environment typically comes with obligations on project managers to do long-term forecasts and plans. Resources must be booked early, because others, who are also interested in using those resources, may book and block them far in advance as well. The project may need certain pieces of equipment or product components quickly, and they must be firmly ordered beforehand to be available when they are needed. Predictability and the need for predictive approaches often come hand in hand. Another factor in predictable projects is often that the projects need a major number of licenses and approvals from government agencies, and any change request would necessitate renewal of these documents, which would be costly and time consuming, such that changes are avoided where possible. One may also consider a project to send a space probe to another planet. During the probe's flight, the changes that can be applied to the project are rather small.

Seven percent of the survey respondents reported that the requirements on their projects were rather static. These are the projects whose success can be measured by "meeting original goals". If one focuses just on this group, the CHAOS Report and similar work mentioned above can have value. These projects allow for long predictions. The project manager and the team have an early understanding of what they need to deliver at the end and how they will achieve the deliverables. They may be in an industry where change is comparatively slow. Traditional chemical engineering is a good example. For example, the fundamental laws of chemistry do not change quickly, new laws take time to be ratified, and basic research to design and erect production facilities follows a predefined process. Things may be different when one moves from traditional chemical engineering to bio-technology because changes happen much quicker. Bridges are another example. Once a bridge design has been approved, the leeway for changes is small to avoid delays and additional costs and from having to obtain new approvals.

Predictable projects commonly require granular planning. A long-term plan needs a deeper level of detail to become accurate and reliable enough to not get derailed by overlooked peculiarities. These projects may also require a top-down approach to ensure that the plan is being implemented. Many readers may know the U.S. TV series *The A Team* of the 1980s and may also remember John "Hannibal" Smith, whose catchphrase, "I love it when a plan comes together", is also the perfect motto for this type of project.

3.11.2 Exploratory Projects

Slightly more than 30% in my survey worked on these types of projects.

The best description for this type of project that I know was written in a poem by Spanish poet Antonio Machado[*]:

> *Caminante no hay camino, se hace camino al andar.*
> Wayfarer, there is no way, the way is made by walking.

A project may be following well-developed paths, or the path may be made by walking; this is what my experts call "exploratory projects". A family of methods was developed for these projects that are more driven by discovery than by following a plan, mostly by people from software development. They looked deeper into this type of project and developed the so-called "agile" approaches that are also used in other industries, including engineering, organizational development, basic research, and creative projects. It is typical that requirements for these projects are identified and formulated during the course of the project by the project team and other stakeholders. Adaptiveness and agility of the teams must be high in these projects, as the preferred way to proceed may not be the best, and the project must then move to another approach. Agile methods are often regarded as a reluctance to plan, but this is a misunderstanding. They stand for a discipline of frequent, short-term planning building on knowledge and results that have been achieved so far.

Exploratory projects are projects in which the teams identify the requirements while the project progresses. The internal requester, external customer, and other stakeholders who mandate the project cannot say in detail what the result should look like, and other aspects of the project, such as delivery time and budget, are also identified during the course of the project.

3.11.3 Projects with Frequently Changing Requirements

Between the two extremes, predictive and exploratory, are projects that have requirements described early, but they are not final and undergo frequent change during the course of the project. Causes for change often can be due to changes in the business or legal environment, new management strategies, or the organization having learned a lesson from another project that should be included in the current one. The term "change request" is often used for scope changes such as additional features that the project needs to deliver. Change requests additionally can be used for many other changes in the project. For example, a functional manager has temporarily provided an employee for the project as a team member and wants the person back. Having a well-defined, documented, and agreed upon change request management procedure may now be helpful. This would enable the request to go through the same disciplined process as scope changes, including impact analysis and rules for communications. It is then just another knowledge area to which it applies—human resource management—and an agreed-upon change request procedure, ideally including a standardized change request form. The change request form would be a filter separating the requests that are important for the requesting stakeholder from the less relevant ones that stakeholders would not consider worth pursuing. Or consider a situation in a customer project, when the internal sponsor, the project

---

[*] (Machado, 2012).

manager's boss, requires costs to be cut by 20%—of course without impacting customer satisfaction—to increase the margin from a project. We remember that a customer project is a profit center. A good way to handle such a request is for the project manager to open the drawer of his or her desk, take the change request form out and ask the sponsor, "Please fill in; we will then assess the impacts". I will go more into detail into this kind of protective change request management in Chapter 5.

What would the project manager do if the functional manager wants a team member to be returned from the project who is considered a team nuisance anyway or an annoying nay-sayer who slows down progress and toxifies the atmosphere? One could respond, "Dear colleague, may I fill in the change request form for you?" A change request form, agreed upon and pronounced mandatory for all change requests early in the project can be a great way to protect a project from damaging change requests and separate them from beneficial ones. Over 50% of the survey attendees reported that their projects undergo frequent change. For these projects, a working change request procedure may make the difference between success and failure. I will go more into detail into this kind of protective change request management in Chapter 5.

The continuum and tension-zone between long-term predictions and short-term agility has been a topic in project management from its first descriptions. An early example was an article in a civil engineering magazine of 1915 from Switzerland. In this article, Hermann Schürch described the construction of a particularly difficult railway bridge in the Alps ("Langwieser Viadukt", an icon for the Swiss as the Golden Gate Bridge is for U.S. Americans) in 1912–1914[*]. Weather changes and ground consistency were among the most important risk factors that influenced whether the project could meet a tight schedule. In addition, delays were caused by approvals that were necessary to get the project started but were not granted on time. Building bridges needs predictability and order, and these were factors that were hard to foresee in this specific project. Schürch reports how a progressive planning approach was used to reflect these uncertainties.

The topic turned up again when a team led by John Fondahl developed a manual network diagramming in the late 1950s. He recalled in his memories of 1987[†] the problem of predictive plans with high granularity, into which a lot of time and effort was invested, and how these efforts became a burden when changes became necessary. The more effort a team has put into planning, the harder it gets to rework these plans to adapt them to changes. He recommended for projects that expect a high number of changes to use a "rolling basis" we call it today "rolling wave". Other terms in use are "progressive elaboration" and "iterative-incremental". The three terms mean essentially the same. They describe a form of project management with a planning horizon of medium duration, longer and more granular than in agile methods used for exploratory projects but shorter and commonly less granular than in the foreseeable world of predictable projects.

As I mentioned earlier, the typology described herein is not only a typology of projects, but also of project situations. It may well be that a project has periods in which predictability is high, and others when it is low. Chapter 4 will discuss approaches that one can take to master these different projects and project situations.

---

[*] (Weaver, 2012).
[†] (Fondahl, 1987).

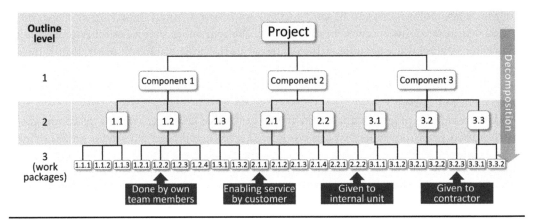

Figure 3-13   The project is decomposed into smaller components.

3.12   Composed Projects and Decomposed Projects

Most literature describes project management based on decomposition: The project manager or project management team develops a Work Breakdown Structure (WBS) top-down to break down a large endeavor that is so hard to plan and manage into smaller, more manageable components. The basic process follows the description shown in Figure 3-13.

The schematic illustration in Figure 3-13 shows the breakdown process, which creates a kind of upside-down tree. Decomposition often follows chronology, the first-level components may then be phases. One can also build the WBS based on organizational allocations, and with this approach, the first-level components would then designate business units involved, or based on function groups of the final deliverables. One can also mix the different breakdown criteria, but my recommendation is to restrict each outline level to one criterion, and then if desired, use another criterion for the next level. The components at the lowest level are called "work packages", and they consist of activities, the individual actions that are performed by team members individually or in small groups[*]. The diagram also shows that work packages may be allocated to internal business units of the performing organization, may be outsourced to external vendors, or may be done by the project team. In the latter case, the project manager and the project management team will have to plan and track the activities that are part of the work package by themselves, while work packages that have been assigned to units or contractors are generally managed by these units or companies. This is the process that is traditionally used to give the project an architecture and ensure complete assignment of work. A WBS is often also used to communicate cost information, following the simple principle that the total cost of the project is the sum of all individual costs; hence the "100% principle" of the WBS. Work items not mentioned in the WBS will not be done, while items that are in the WBS are part of the project. This approach can be used in internal project context to ensure a common understanding, or in a customer project, where a WBS may be part of a contract.

---

[*] In project management software, WBS components on all levels and activities are not treated separately and are then commonly called "tasks". The difference between WBS elements and activities is an older approach and is easier for manual planning and tracking of projects.

There are projects that are created using a bottom-up approach. Individuals, units, or organizational leaders voluntarily come together to jointly contribute to a project that they all believe should be pursued. Their interest may be driven by the emergence of a business opportunity, and they prepare a business case that they can better exploit by acting together. They may also have some shared ethical or political ambitions or expect a greater advance in technology and want to be actively involved. Work packages, activities, and other kinds of tasks in a composed project are not assigned; instead, they are rather self-selected. The typological dimension came up once in the research, when a field expert considered it a root cause for failure of certain practices that were successful in other projects. In this case, a project manager had assumed that he could manage a composed project with the same tools and behaviors that he had used before to manage decomposed projects, and he failed. It was his understanding that he could assign work when the team members understood themselves as volunteers that led to avoidable difficulties with the consequence of massive team attrition.

When I talked with other experts about this type of project, a common reaction was that they may exist but are rare. As I believe in such cases that it may be better to talk with people than about them, I asked practitioners from project management in an ad hoc survey; Figure 3-14 shows the results.

The respondents reported that decomposed projects are more abundant, but a ratio of 30% of all reported projects is far from being rare. The number may grow over time. Agile methods, described in more detail in the next chapter, generally prefer to compose projects from individual contributions. There are some companies today that follow a principle called "Leaderless Organizations". The basic ideas were laid out in a 2006 book by Ori Brafman and Rod A.

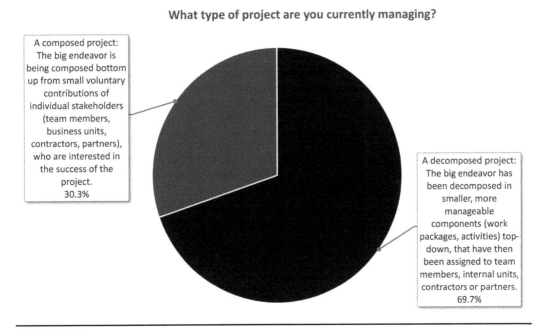

What type of project are you currently managing?

A composed project: The big endeavor is being composed bottom up from small voluntary contributions of individual stakeholders (team members, business units, contractors, partners), who are interested in the success of the project. 30.3%

A decomposed project: The big endeavor has been decomposed in smaller, more manageable components (work packages, activities) top-down, that have then been assigned to team members, internal units, contractors or partners. 69.7%

Figure 3-14   Composed and decomposed projects.

Beckstrom titled *The Starfish and the Spider*[*]. The metaphor compares the spider, whose neuronal system is centralized, which is crippled if it loses a leg and dies if someone cuts off the head, and the decentralized starfish, which can stay alive and even regrow a lost leg, and the lost leg can grow to become a new starfish. Some leaderless organizations are highly successful, such as the U.S.-based manufacturer of special textiles WL Gore (catchphrase: "No Ranks, No Titles"), or Valve Corporation, the makers of video games and supporting services and systems. Setting up a leaderless organization can be difficult, as the shoe distributor Zappos had learned in 2015, when about 14% of their employees left the company rather than support the change[†], and a customer of mine who made such a change some years ago also had a similar experience. What I observed is that there were problems with certain employees who felt deprived of a boss telling them what to do and saying "well done" when they were finished. Striving for higher goods for their communities, many not-for-profit organizations are actually leaderless organizations, but so are the sleeper networks of Al Qaida, the Ku Klux Klan, and other establishments of organized crime.

Composed projects are temporary leaderless organizations and cannot be managed in a top-down way. Sometimes, they are performed without any formal project managers at all. If they have project managers, they commonly form a small team rather than letting a single person do the job, and their task is more coordinating, moderating, and facilitating than leading in a sense of command and control. Tasks are rarely assigned to team members, business units, or contractors; they are instead voluntarily selected by those contributors, who like to do them, with the risk that unattractive tasks may not be completed for some time, while others may be done concurrently by teams duplicating the same work. Contributors delivering poor work are often another dilemma. Telling them the truth about their products may destroy their intrinsic motivation as contributors, but not telling them may lead to quality problems.

A composed project can be perceived as a kind of common property by its contributors (see Chapter 2) and suffer from the "Tragedy of the Commons", when resources are exploited that should be treated with care and preserved. It can also suffer from sensitivities. The volunteers in composed projects contribute with a lot of passion and positive emotion—they would not contribute if these did not exist. This also makes them vulnerable to perceptions of being disregarded or even bullied, feelings that may be unfounded or have grown from misunderstandings, but they can nevertheless be strong and highly damaging to the project. A good rule in a composed project is therefore to praise high-performing team members loudly but criticize poor performance in private.

The next case story shows another risk, which comes with voluntary investments that contributors need to make. Desertec was a composed project (rather a program) started by the Desertec Foundation and the DII industrial consortium in 2009. It was started to support the development of renewable energy production, mainly solar and wind, in arid countries as well as to transport solutions to developed countries by a grid of high-voltage direct current (DC, like from a battery but at much higher voltages) lines. The focus was North Africa as a producing country and Europe as the consuming one, but this focus blurred, and Desertec is also active in Latin America and China. While Desertec was effective in supporting technological progress in the field, it lost important contributors over time, especially the German corporations Bosch and Siemens, which capitulated to global competition and abandoned their solar

---

[*] (Brafman & Beckstrom, 2006).

[†] (McGregor, 2015).

energy business. At the same time, North African states reconsidered their business case and concluded they would rather use the ample energy that can be produced in the Sahara desert to develop their own industry. Changes in business strategies of contributors may impact composed projects as much as the feeling by some contributors that there is an imbalance between their investments and the returns that they receive on them. A local loss of limbs can be overcome by the self-healing abilities of composed projects, as the starfish metaphor vividly describes, but a massive loss of contributors will leave the project without resources. As much as the formation of the project can be a result of group enthusiasm and swarm intelligence, its dissolution may be due to group disappointment or swarm imbecility. Desertec is still alive, but much less focused, and instead is trying to find tasks it can perform using its resources rather than following a specific vision.

3.13 Further Types of Projects

The typology described above has some benefits. It was developed following a published process, which can be repeated by others possibly confirming the results, or adding new insights. It is considered an open typology, which means that it can be expanded when other types of projects are identified, just like Linnè's classification system for species in biology is open for new species or the periodic table in chemistry for new elements. The nine dimensions described so far are not complete; they are the ones that emerged during the research, and there are others that wait for their detailed identification and description.

From my observations as a practitioner and as a trainer, the following come to mind, which I will describe in the following sections briefly.

3.13.1 Engineers' Projects and Gardeners' Projects

One of the difficult businesses to develop self-sustainability and economic health is the winery business. When one has found a good piece of land to grow grapes, one has to prepare the land and plant the young seedlings. The plants will grow the first economically useful grapes in the third year. With good care, the yield will grow over the next years in volume and quality. It may take ten years to turn the vineyard into a profitable business.

Project management's origins are from engineering, and for many project managers, there is only one value development curve that they know: the value of their deliverables are highest during handover and will then depreciate because of wear and tear, aging, outdating, and other causes. People who are responsible for the deliverables may slow down the process by careful treatment, regular maintenance and updates from time to time, but they will not be able to stop the process. Gardening projects are different since the value of the deliverables is expected to grow after handover, and it may take years for the deliverables develop their full value. See Figure 3-15 for the different value streams in the two project types.

Operating systems for modern mobile devices such as Apple's iOS or Google's Android are another example for Gardeners' projects. When they were launched, they had of course the value that they could help run the phone, as operating systems. iOS and Android had an additional purpose, as they allowed third-party software developers to create additional software, so-called "apps", and each app increased the value of the operating systems. The idea was on

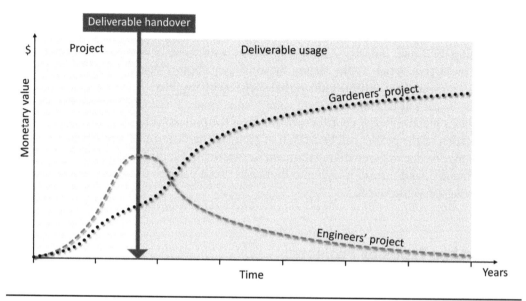

Figure 3-15   The differences in lifecycle value creation of engineers' and gardeners' projects.

the market before, but Apple and Google came into market when the market window opened. They were the first to come up with a complete "ecosystem", which included online stores where apps could be downloaded, combined with capacitive screens that allowed "swiping" and other finger gestures in an easy-to-learn and useful alternative to mini keyboards and styli (a stylus was a small plastic peg to write on a display as if it were paper and pencil). While the value of Apple and Google went up with the value of the operating system, Nokia's value went down, since it became outdated much faster than what they had expected, and when they came with similar products in response, it was too late and the market window had closed again. In the example, engineers lost the competition against the gardeners.

3.13.2 Discretionary Projects and Mandatory Projects

These are two subtypes of internal projects. Unless a company would be forced to run a project under contract for another organization, customer projects are generally discretionary. The decision by the contractor is made during the bid/no-bid decision moment, and possibly again when the customer has accepted the vendor's bid or proposal.

Discretionary projects have a business case or another aspect that makes them attractive for the performing organization. The business case is often a tangible benefit that the organization wishes to obtain by the project, such as additional earnings, cost savings in operations, or improved strategic position. Some readers may be familiar with project selection methods such as strategic scoring, net present value, or internal rate of return; discretionary projects are those where these methods would be applied.

Mandatory projects in contrast are required by law or are necessary to avoid an emerging business crisis. Their purpose may also be to master such a crisis if it has already occurred. Mandatory projects may be deliberate responses to compliance rules, but they may also be done in sheer panic, especially if they were started too late, and law has set firm deadlines that are enforced with severe penalties but difficult to meet. While internal projects are commonly performed in a weak matrix, which means project managers have difficulties to obtain resources for their projects, these projects generally enjoy a strong matrix, and the project manager is generally expected to obtain tangible support from the functional organization.

3.13.3 Single-Handover Projects and Multiple-Handover Projects

Although I discussed this topic in Chapter 1, I consider it important enough to look at it from the typological perspective. Another classical assumption is that a project has a single handover, which ends the project (the team may have to finish some administrative items but will not have additional work to do) and starts the operational use of the deliverables by an internal recipient, e.g., a functional department. It is interesting to see that many projects across different industries and application areas have changed to multiple handovers, also referred to as "staged deliveries", as shown in Figure 3-16.

The deliverables from these projects are not transferred as one massive piece but in portions during the course of the project. Multiple handovers can be a solution in a situation of time pressure: work on a project will not be able to be completed by an imposed date, so prioritization is done: What is most important? What is mature enough to be finished on time and then delivered? Multiple handovers may also be part of a proactive organizational project management strategy of providing quick wins to the requester or customer and allowing the use of a partially finished product in order to gain the first benefits early. In addition, they can build trust in the project team and observe how users work with the deliverables, how the deliverables perform, and allow one to make adjustments for further development where necessary. Staged deliveries in this strategic understanding are also called evolutionary deliveries, and there is even an extreme form called "continuous delivery"*. This latter approach is mostly used for software development with the intention to have frequent handovers of small

---

* Described in Detail in Jez Humble and David Farley's book *Continuous Delivery* (Humble & Farley, 2010).

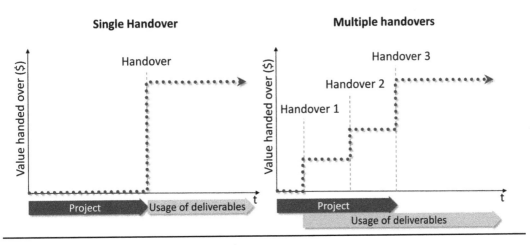

Figure 3-16   Multiple handovers deliver value in stages.

increments of the software to its users. Some users will find the approach fantastic, as it ensures tight interaction between them and the developers; others may be terrified by the expectancy of the need to learn new software functions every other week and the outlook that functions one has understood may be changed again. One obviously has to use these approaches with a sense of situational proportion and empathy for people involved. Multiple handovers influence many other aspects of project management, including investment calculations, so they need to be taken seriously. The return on investment begins before the entire investment has been finished, which makes adjustments necessary to the mathematical models used.

3.13.4 No-Deadline Projects, Single-Deadline Projects, and Multiple-Deadline Projects

In literature on project management, but also in training, a common presumption is that a project has a deadline that it must meet. This presumption often comes together with the previous one of the single handover or delivery. Again, I wondered to what degree this presumption correctly describes the situation that project managers are experiencing, and I asked them in another ad hoc survey between September 16 and 18, 2015. I received 402 responses by individuals who are managing a project with results shown in Figure 3-17.

Forty percent, the largest group of respondents, said that their projects have not just one deadline but several of them. The second group (a bit more than one-third of the respondents) met the presumption of one deadline. Almost one-fourth of the respondents stated that their project does not have a deadline, which does not necessarily mean that the project is not performed under time pressure.

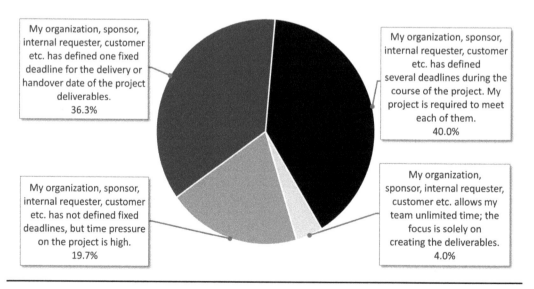

Regarding time pressure, what statement is true for your current project?

My organization, sponsor, internal requester, customer etc. has defined one fixed deadline for the delivery or handover date of the project deliverables.
36.3%

My organization, sponsor, internal requester, customer etc. has defined several deadlines during the course of the project. My project is required to meet each of them.
40.0%

My organization, sponsor, internal requester, customer etc. has not defined fixed deadlines, but time pressure on the project is high.
19.7%

My organization, sponsor, internal requester, customer etc. allows my team unlimited time; the focus is solely on creating the deliverables.
4.0%

Figure 3-17   How project managers reported their perceptions on time pressure and deadlines.

There are several reasons why a project may have multiple deadlines, and survey participants used the comment opportunity that they had in the survey to talk about them. One wrote, "as a side note, most of my projects come with a fixed timeline". A second wrote, "Many of my clients fix 'arbitrary' deadlines without proper planning". I found such situations typical for many industries: Procurement managers send out Requests for Proposals (RFPs) or Invitations for Bid (IFBs) with Statements of Work (SOWs, mostly narrative requirement descriptions) attached that already include a timeline, which consists of a series of fixed deadlines, often falsely referred to as milestones, and when vendors make their decisions to send proposals or bids in response, they accept the deadlines, often without having given a project manager the opportunity to validate these dates for feasibility. A project manager doing such an assessment could also forecast the consequences for the vendor company that receives the award, but the offer is often submitted to the prospect without such a review. When the contract is the awarded by the customer, the contractor will have to find a project manager, assign the person, and the project manager will then have to meet the deadlines that have never been checked for plausibility. Another respondent explained the situation with a "classic stage gate process". A stage gate or phase gate is a review that is placed between two sequential phases to validate the correct performance of the previous and closed phase and obtain the approval to enter the next one. During a classical stage gate, no work is done, and the only active process is the review. It is common practice in many industries that the end of a phase is linked with a deadline to ensure timely beginning of the gate review. Project managers have to ensure that deadlines are met, so these different deadline situations pose different requirements on the practices used by project managers.

3.13.5 One-Shot Projects vs. Multi-Shot Projects

Have you ever noticed that around 2010, the packages of cellphones have generally become smaller, while at the same time, after years of constant shrinkage to an almost unhandy minimum somewhere near to lipstick cases around the year 2008, the actual phones, especially smartphones, have become bigger again? I think there is an interesting story behind this trend, and yes, it has a lot to do with project management.

During unboxing of an expensive item, a package that is considerably larger than its content can give an upgrading experience to the product inside. Makers of posh perfumes know and use this effect well. The white (or sometimes colored) space around a small flacon in a box creates an impression of luxury and enhances the value perception linked with the fragrance, comparable to a *passe-partout* of a picture in a frame. Books with poems or glossy photos use the same effect. A wealth of whitespace around the content celebrates just this content, and while actually being utterly useless, enhances its visual appeal.

Handset makers have also used this effect. When the customer opened the lid of the box, the phone inside was found resting in some kind of bedding, often similar to that around a bottle of *Eau de Parfum,* obviously intended to make it look prestigious and increase the experience of joy and pleasure that the user expected from the technical features of the product. Today, many cellphone makers cannot afford this luxury. Why? I began to understand this better after I read an article at Bloomberg.com by Adam Satariano[*]. It deals with the logistics

---

[*] (Satariano, 2013).

of the launch of the Apple iPhone 5S and 5C and with manufacturing and shipping of the first batches that were made available for customers on September 20, 2013.

The iPhone contributes to Apple's revenues significantly, and during the immediate weeks after product release of a new type, customers are prepared to pay more attention to the product and also to pay extra money for it; a short-term, high price/high volume opportunity that Apple definitively does not want to miss. It has to make sure that both the phones and the phone customers come together in the shops at this point in time. Ensuring a successful product launch is not an operational task. It is a true project, a temporary, unique, and a thorny endeavor.

An example how a company may fail in creating this kind of customer experience is that of Google after the launch of the Google LG Nexus 4 smartphone in late 2012. While the product enjoyed high customer demand, the Korean company LG, the manufacturer for Google, was not sufficiently prepared and could not make and deliver enough phones. Google as the sales point had to communicate delivery times of up to nine weeks—occasionally, it did not communicate a delivery time at all. Customers who were looking for excitement and wanted to be among the first to show off the phone to their friends faced frustration instead.

An example of a reverse miscalculation was Microsoft, who had to re-assess the value of its stocks in tablet computers called "Surface RT" after a $150 price decrease in 2013. It had its newly launched tablet placed in shops and stored in warehouses in sufficient quantities, but it lacked customers. In order to get these "channel stocks" reduced, it then tried to lure buyers by dropping prices by up to 30%. The follow-up, the "Surface 2" tablet, was about to be launched, and Microsoft did not want sales channels to be blocked by the old model.

Coming back to the logistics project at Apple, on top of the risk of having an imbalance between item stocks and clients, there is another risk: Having a surplus of phones in a place with low demand and at the same time a shortage somewhere else. And to make the situation even more difficult, the two new types of iPhones come in two models with different configurations regarding colors, gigabytes of memory built-in, and adaptations to five different international mobile network standards that the phones need to support. In total, Apple had to manage 40 different versions whose sales needed to be predicted. Apple did not want to have misplaced numbers, types, and configurations in warehouses and stores. Instead, it wanted to make sure that customers in all markets find just those iPhones that people are prepared to pay for and which they can use in their home networks—otherwise, they would send the phones back to Apple in numbers—a nightmare for every vendor.

Apple had only one shot. It did not want to miss the extra revenues and positive market attention that it could get during the after-launch sales peak, which may last one or two months and would then ebb down over some time to finally become normal sales figures. Another point of concern for Apple: Once the release date—September 20, 2013—was announced, it could no longer be postponed without damaging the company's reputation on the various markets. The launch date was then a highly inflexible deadline.

The project (rather a program) at Apple to prepare for the product launch must have been started months before the actual launch. It included running development projects for hardware and software but also for packaging, marketing, and initial logistics in a concurrent and coordinated fashion. It definitively included legal research—Apple did not want the launch to be thwarted by injunctions from competitors or patent trolls. This program was not disintegrated into "profit centers", instead, it was obviously managed in a highly integrative fashion.

It also included developing resources for the processing of sales and logistics data to enable quick decisions on where to direct delivery batches of phones, and also on the order and quantities of the next configurations for production.

Transportation by ships would have been inexpensive, but it would also have been slow and did not allow for the flexibility that Apple wanted to design into the launch process. Apple could also have used regular airfreight transportation services, but they could not meet its demands regarding speed, flexibility, and control over the shipment. Apple chartered planes. Big freight planes.

Satariano reports that Apple could pack about 450,000 iPhones into the body of a Boeing 777, which cost about $242,000 to charter. And this is where the squeezy packaging issue appears. For such a business model, Apple could not afford redundant space around the iPhone in the box. This would have immediately reduced the number of items that can be packed together in an aircraft, slow down transportation, and increase cost. So, Apple optimized the entire process around a bottleneck, which was the aircraft body. This way, it added speed and flexibility to Apple's logistics project while ensuring economic viability.

I consider the launch of the iPhone 5C and 5S models a perfect example of a one-shot project. If you blow it, you will not get a second chance. In these projects, the project manager needs a realistic and viable plan. One has to create predictability and must plan for the things one knows as well as for the foreseeable uncertainties. And one has to anticipate the relevant bottlenecks and develop the entire project around them.

There are other projects where the project manager and his or her team have several shots. They can use staged deliveries and handovers of partially completed products to develop their final product in a step-by-step approach and fix errors as they occur. They can easily allow for change requests, as long as they are given the resources and time to manage them in a coordinated and controlled fashion.

One-shot projects can be found in many application areas of project management: A major meeting or congress may be a popular example, and an election campaign is another one. When the election is over, one can no longer fix errors in the project.

One-shot projects have an increased risk of failure. The already mentioned Terminal 5 at Heathrow Airport is a famous example of a project that completely failed immediately after its hand-over—actually a grand opening performed by the Queen of England—in March 2008, and it took the sponsoring organizations British Airport Authorities and British Airways weeks to become fully operational again and months to recover from the damages occurred. An example of a successful one-shot project was the project to relocate Münich Airport in 1992, which was moved from one location to a new one overnight and started operations there again on the next morning. Proactive risk management is an essential requirement for them. When I pointed out the insufficient ramp-up plan for the terminal, it deals with having reserves in place. Limiting the number of flights would have left some personnel and equipment capacity free to fix problems as they occur, and the service then could have been ramped up to full productivity at a rate of increase based on the growing trust in the functioning of the system's terminal operations. It would have saved costs and a major embarrassment.

Obviously, Apple was dedicated that it was not going to fail during the launch of its new products. This is a promise that can be kept only with professional project management and dedicated staff. An educated guess would be that the project/program was run inside a very strong matrix, and the project managers did not suffer from lack of management attention or other resources. The entire program was highly integrated; always bearing in mind that one failed project would jeopardize the success of the entire program.

Situationally sound project management is a great basis for success: Apple (2013) communicated in a press release that the firm sold a record-breaking number of nine million handsets in the first three days after the launch.

Apple's CEO, Tim Crook, said that "the demand for the new iPhones has been incredible, and while we've sold out of our initial supply of iPhone 5s, stores continue to receive new iPhone shipments regularly".

And what about this special unpacking appeal that customers may miss when Apple sends its phones in rather small and cramped boxes? Apple provides remedies for doing so. The first one is probably the name, which stands for success and style in the eyes of many customers. Product quality is the second one. A third remedy may be the feeling to belong to a loyal elite group of smartphone users around the world. Apple may also think that the unpacking experience does not matter that much: When the customer opens the box with the phone, the critical decision to buy it has already been made.

Chapter 4

Practices for SitPM

Several sources describe so-called project life cycles to select from. The use of the word in this context can be misunderstood. It is not used to describe phases that the project will inevitably go through, but is used instead to explain approaches to managing a project. In Chapter 3, I described project types, which means a typification to help understand what specific projects are. In this chapter, several topics may seem to be repeated, especially the tension field between predictable and Agile projects. There is nevertheless a fundamental difference, because in this chapter, the question is not what the project is, but how you approach and manage it.

In Chapter 3, you were an observer, trying to understand; now, you become an actor. When you are assigned to a project, you have only a few options to select from: yes, no, and maybe a "yes, but . . .". Often, there is not an option to say no. After that moment of assignment, you are responsible for how you manage the project, your decisions and actions, as well as their outcomes, intended and unintended. There are of course some "Zombie" projects, in which project managers have no chance whatsoever to be successful. It is mainly the beginners in project management who are assigned to these Zombie projects; people who still lack the experience that with all the effort and acumen that they will invest, they are bound for failure.

This chapter relates to all other projects, in which project managers have a chance to meet their duties and have great projects. You need to understand which practices can lead to success and which ones cannot. Robert Wysocki[*] used a four-quadrant model to describe lifecycle approaches, see Figure 4-1.

The dotted grey arrow describes how, in his understanding, uncertainty grows from the bottom left quadrant to the top right. The approaches that he recommends depend on the quadrant in which a certain project is located:

- **Traditional.** Well-defined projects, in which the project goal is specified and the solution to achieve this goal is also known—in most cases, because the solution can be derived from other, similar projects. This is the classic concept of a highly predictable project.

---

[*] (Wysocki, 2014a, p. 8).

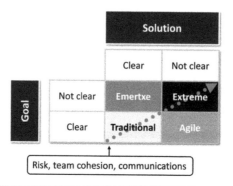

Figure 4-1 Wysocki's four-quadrant model.

- **Agile.** The goal is clear, but the solution that leads to achieving the goal must be developed during the course of the project. An organization may need a new machine or software to meet specified objectives.

- **Emertxe*.** The solution is clear, but the goal needs to be discovered during the course of the project. The organization's leaders have decided to implement a new machine or a new software and explore new opportunities that come with the solution.

- **Extreme.** Neither the solution nor the goal are known. The old machine or software is outdated, and all the organization's leaders know at this point in time is that a replacement is needed.

The interesting thing about Wysocki's work is that he not only rejects a "one-size-fits-all" understanding of project management but also gives advice on the specific use of practices. While his approach focuses on life cycles and process assets[†], I am more interested in behaviors and approaches. The two points of view are not contradictory, but complement each other. The focus in the following discussion will rather be on the dotted line in Figure 4-1, which describes the tension field between predictable projects and total uncertainty. It also stands for the approach that the team must take to cope with the growing uncertainty by increasing team cohesion and intensifying communications when risks are increasing.

Another aspect of the chapter will be leadership behaviors. I will use a model by Professor Jean Lipman-Blumen, called the Achieving Styles Model, which uses three groups of nine styles to describe how people in charge master their challenges. For the approaches and the achieving styles, I will give some hints how they can be favorable or detrimental in projects or project situations of a given type.

---

[*] The word was coined by reversing the word "extreme".

[†] Robert Wysocki describes a situational project life cycle in another book *Effective Complex Project Management*. It begins with an Agile setup phase and ends in a more predictive execution phase. This lifecycle, called Adaptive Complex Project Framework (ACPF), will probably be appropriate for many projects (Wysocki, 2014b).

4.1 Introductory Questions

This chapter discusses some more topics that may be new to most readers. Some of the following introductory questions relate to these topics and may therefore be hard to answer. They require knowledge that readers will first need to obtain by reading and probably also by reflecting on their own behaviors and those of people they work with and have worked with. It may also be interesting to research your own experience and consider observations relating to the performance of others to try to find both confirmations and contradictions.

You may try the questions now, or read the chapter and then answer them.

---

1. A corporate Project Management Office (PMO), which governs a major number of internal projects of an organization, made the decision to change to Agile methods for its entire project portfolio. What are the detrimental effects of this decision?

   a) Agile methods require long-term planning that may not be possible to implement for all projects in the portfolio.
   b) Certain projects or situations during projects may require more predictive approaches that Agile methods do not sufficiently support.
   c) The training demand for Agile methods is massive, and may block resources that the organization needs to meet their contractual obligations.
   d) Agile methods consist of a strong top-down approach that employees may consider outdated and deeply demotivating.

---

2. Which is a common signal that a project manager applies the "social" achieving style?

   a) The leader supports others from "behind the scenes".
   b) The leader creates an aura of "charisma" around him or her.
   c) The leader delegates tasks to the person who is the best fit.
   d) The leader delegates tasks as "stretch assignments".

---

3. "Iterative-incremental" is another word for what kind of project management approach?

   a) Rolling wave
   b) Waterfall
   c) Agile
   d) Predictive

---

4. A project manager acts as the most productive team member and subject matter expert with the highest degree of expertise. Which achieving style does the person apply?

   a) Social
   b) Power
   c) Intrinsic
   d) Vicarious

---

5. If "the way is made by walking" in a project, which approach should be considered the most promising one?

   a) Waterfall
   b) Agile
   c) Top-down
   d) Rolling-wave

6. A customer project is going to be performed for a client organization and the project manager has just been chartered by the contractor organization.

The manager in charge on the customer side has a history of former projects during which interests of the organization were enforced against the various contractors in a highly forceful manner. Contractors often finished their projects with losses, and several of them were later sued for minor mistakes that their teams made during the projects. Which achieving style best represents the manager on the client side so the project manager is prepared?

a) Competitive
b) Collaborative
c) Entrusting
d) Power

4.2 Lifecycle Approaches

Several sources describe so-called project lifecycles. They actually mean approaches to project management; it may well be that these approaches need to be adjusted during the course of the project to different tasks and situations. A project to build a bridge near to a cultural heritage location may have an early creative phase that is dedicated to developing a design that avoids disruption of the historic view. Creative phases are commonly better managed with Agile approaches, as creativity is generally hard to predict. Later in the same project, when the bridge is physically built, a predictive approach will be necessary. Resources must be booked early, work must be procured beforehand to ensure that the results will be available on time, and the project cannot have many changes that would necessitate the renewal of existing approvals by government agencies and other stakeholders.

In this book, I use the term "lifecycle approaches" to reconcile the two definitions. This should not be misinterpreted that they are necessarily used identically from the beginning to the end of a project.

4.3 Agile Approaches

In a discussion with the head of the Project Management Office (PMO) of a company from the electronics business, I was told in 2014 that the company was switching all projects to "Agile methods". Agile methods were popular between 2000 and 2010 and are often considered another form of "best practices" by some companies and consultants. Others made the change to Agile methods for their projects with high hopes and were disappointed to some degree. Talking with experts in Agile methods, their common response is that the methods have been applied incompletely, and that too many projects have the term Agile as a label but are still "Waterfall" and "command and control" projects. For some people, agility has become an ideology or a dogma, something that I call "Agilism". I think that it is time to explain the terms, as not all readers may be familiar with all of them.

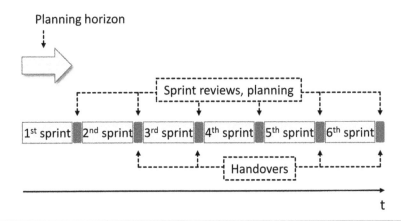

Figure 4-2 A project performed along a sequence of sprints with fixed time boxes.

Agile methods expect that change requests will come quickly and frequently, and they should be generally accepted. As each of these changes will come unexpectedly, there is no firm long-term planning; instead, planning is limited to one to four weeks' development cycles, often referred to as "sprints". It is a misconception that there is no planning in Agile methods since they require frequent planning at the beginning of each sprint. Sprint durations are fixed—an approach also called "time boxing". During a sprint, no changes are implemented; they are collected and then discussed and decided upon between the sprints. At the end of a sprint, the team also discusses what it has achieved so far and bases decisions for the next sprint on the results (see Figure 4-2).

Agile approaches are commonly based on a document called the "Manifesto for Agile Software Development", often just referred to as the "Agile Manifesto"[*]. Starting in software development, the concepts were transferred to other applications for projects. There is, for instance, an "Agile organization development manifesto" for projects dealing with organizational development. The principles have been formalized in several methodological approaches, of which a method called "Scrum" has become the most popular. The basic assumption in Agile methods is that internal requestors of projects, customers, users of deliverables, and so on are generally unable to describe their requirements against these deliverables with sufficient accuracy and completeness for a team to just develop them. This assumption leads to constant change that is not manageable by a project team. For this reason, Scrum uses so-called sprints (periods with fixed durations commonly between one and four weeks), during which no change requests are brought up to the team. Change requests are instead collected by a role called the "product owner". Between two sprints, there is some time to review and discuss the results of the sprints, and during this time, the change requests can be brought to the team for implementation in one of the next sprints, unless they are rejected. Scrum has another role, called the "Scrum master", who ensures that the team adheres to Scrum principles but is not a project manager; the Scrum master is more of a moderator position. This interplay of periods with static requirements and moments when changes are implemented has some benefits. One is that the team develops a working rhythm, which can make it quite productive. Another is

---

[*] (Beck, et al., 2001).

the desire to have a "potentially shippable product" at the end of each sprint, which means that it has the maturity and quality to form a product on its own, something that would be called a "quick win" in other methods. It has some similarity to the work breakdown structure (WBS) used in other methods that has been traditionally structured by deliverables*. Most Scrum teams would nevertheless reject a WBS as too predictive and "command and control" oriented but would follow the principle that their sprints should have a product at the end. I will describe Scrum in more detail in Chapter 5.

Agile methods have a major disadvantage, which is the negligence of work flow and the logical relationships of work items like work packages or activities with the possible conse-quence of rework. I recommend a definition of rework in project management as laying hands again on a deliverable that was considered finished previously in order to repair or alter it. Agile methods mostly reject defining work flows in network diagrams and similar tools as too predictive, restrictive, and command-and-control oriented. They prefer that team members self-organize and also self-select their tasks instead of having them assigned by a project man-ager. Many Agile methods, therefore, even reject having a project manager. This reliance on self-organization can work well, but it can also lead to a kind of "biscuit plate" project manage-ment, which allows team members to select the specific tasks that they like to work on at the moment, similar to how children select biscuits from a dish: "I want the one with chocolate". Tasks performed in the wrong order inevitably lead to rework and are among the top reasons that projects miss deadlines and budgets. The hours it takes to rework a project is a great met-ric to measure the efficiency and quality of work done by a project team. A solution in Agile teams is to never consider a deliverable finished, but this solution collides with the concept of "potentially shippable products", and with the simple fact that at one point in time, the entire project with all its deliverables should be considered finished.

Another problem in practice is that Agile methods work best in solid, well-rehearsed, and dedicated teams comprising subject-matter experts with strong mutual respect. This builds that "bands of brothers and sisters" and actively avoids the emergence of game-theoretical dilemma situations. Colocation, placing the team in one location, is definitively helpful, but distribut-ing a team that has been collocated in the past and successfully went through the often dif-ficult phases of team development can also be a successful approach, as discussed previously. A survey conducted by the Scrum Alliance from 2015 stated that the average team size is seven people, which complies with recommendations made by Scrum experts that a team size of four to nine people provides the best results and is therefore recommended. Many organizations expect their teams to do a lot of multitasking, and they construct teams from individuals who have not gone through the phases of team building and do not have time and physical prox-imity to do this during the project. The team size is also not feasible for all projects, at least not without the risk of team fragmentation. Agile methods seem to work much worse in such environments. On the other hand, having a small team without fragmentation that is well-rehearsed and competent in the subject matter, located in physical proximity, and protected from disruptions from the need to meet other business demands is a good recipe for a success-ful project anyway, regardless of the practice in use.

---

* The *MIL Handbook 881A* that was required for use in the United States Military stated that no ele-ments should be included in a WBS that are not products. The document was superseded in 2011 by MIL-Std. 881C, which no longer has such a requirement (DoD, 2011).

Agile methods are highly disciplined; however, they are often abused as cheap excuses: "We do not have a plan. We are Agile, you know".

I already mentioned Agilism. Some promoters of Agile methods developed some kind of ideology around these methods. They consider Agile approaches and methods generally superior, independent of the specific project situation, and insist on the implementation of aspects of Agile methods (such as self-organization and self-assignment of team members to tasks) with an almost religious zeal. For them, the world of project management consists only of two approaches: Waterfall and Agile. The perseverance for Agile methods comes with a fundamental rejection of the waterfall. Grey-shades between the extremes are not considered. If a person's only tool to earn a living is a hammer, the person must convince people that the world is made of nails.

4.4 Waterfall Approaches

In a waterfall, water can move only in one direction: down. This metaphor is used to describe predictive models with long planning horizon, as shown in Figure 4-3.

Predictive methods are a response to the dilemma situation of the learning process discussed in the second chapter. Early in the project, project managers have many options for decision making, and most are inexpensive, but the knowledge is still not created to see the need for the decisions and to make them. Later in the project, there is much more knowledge, and the need and the options for decision making are much clearer, but the number of these options goes down and becomes more and more expensive. Chapter 2 described how project managers try to bend the learning curve to recognize the needs for decisions and the available options earlier, at a time when these options are more numerous and less expensive. Predictive methods are used as tools that project managers can use to accelerate their learning processes. At their core is network diagramming.

Predictive approaches were used when modern project management began in the late 1950s. Some major endeavors made it necessary to improve planning on top of bar charts and WBS that already existed*. It is interesting that several developments were made at roughly the same time and independent of each other:

---

* The members' magazine of PMI®, *PM Network,* published an article on the memories of that time (Snyder, 1987). Thanks to Jim Snyder, who gave me some additional information in a personal setting.

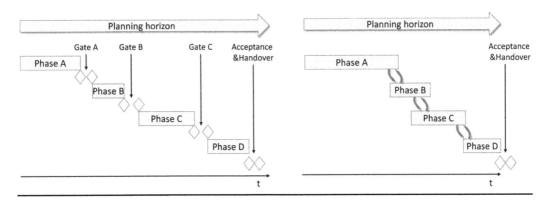

Figure 4-3   Two "waterfalls": A sequential and an overlapping model.

- The Polaris/Fleet Ballistic Missiles program to develop and build submarines and nuclear missiles to be launched from these submarines, programs with unprecedented complexity, on a technical level as much as on organizational level and risk. A small mistake could have sent the world into a third World War. The U.S. Navy had two development mandates for network diagramming given to teams. One was headed by Willard Fazar, which resulted in a method for use with computers named Program Evaluation and Review Technique (PERT)[*]. The other one was led by John Fondahl for manual application, and it led to a method called Precedence Diagramming Method (PDM)[†].
- Developments by Du Pont and Remington Rand Univac, performed by James Kelley and Morgan Walker, which led to the development of a third method called Critical Path Method (CPM)[‡].
- A fourth method was developed in France by Bernard Roy at about the same time, called Metra Potential Method (MPM), for the development and manufacture of the cruise liner, *France,* whose keel was laid down in 1957. When the ship was taken into service in 1962, the *France* was the longest passenger ship ever built, with a length of 315 m (1,035 ft). It incorporated new, component-based production methods that made more sophisticated planning methods necessary to coordinate contractors with long-term assignments. MPM had basic similarities with PDM but used different types of dependencies between activities, which made the diagrams more flexible and meaningful but harder to understand.

In about 1960, PDM was merged with elements of CPM, and in the following years, the flexibility of dependencies between activities that MPM had developed was introduced to PDM. This laid the foundation for the subsequent success of network diagramming to make projects more predictive and help identify problems early, especially deadlines that the project is going to miss, and as a tool to visualize project work flows, simulation, and optimization. A brief introduction on this enhanced PDM network diagramming is provided in Chapter 5.

There are several factors that make long predictions necessary, among them early booking of resources and early awards of contracts to subcontractors who need to have their work results finished on time. Another factor not yet discussed is the need in matrix organizations to coordinate the functional organization and the various projects. Most functional organizations are set up for operations and take from this direction the benefit of better predictability of their work that is due to its repetitiveness. In matrix organizations, there are many points of intersection between the functional business and the projects: They share and exchange resources and deliverables and compete for management attention. When a project makes a major mistake, this impacts the functional organization and vice versa. At these points of intersection a kind of domino effect can take place, when the delay of a project causes delays in the functional organization, and risks, problems, and conflicts are also transferred over the domain border between projects and the line organization. As said before, in most organizations, it is the job of these line organizations with their operations to earn income for the entire organization. Many organizations, therefore, try to force projects into the same

---

[*] (Fazar, 1959).
[†] (Fondahl, 1987).
[‡] (Kelley and Walker, 1989).

predictability that has been so successful in operations since the beginning of the industrial revolution and even possibly much longer. In such cases, the functional organization does not understand the specific element of uncertainties in projects, and often enough, its managers do not want to know it.

The key to predictions are the various network diagramming methods. The PDM method is used by most modern desktop software programs. So-called "Gantt chart" views look quite different, but the underlying logics and mathematics are the same. Further enhancements came later with methods like Graphical Evaluation and Review Technique (GERT), System dynamics modeling[*], and the Critical Chain Project Management (CCPM)[†]. The last one is interesting as it changed some paradigms in network diagramming. Traditionally, the concept was to do work as early as possible, but CCPM recommends to do work just in time by adding dedicated time buffers when critical activities pick up deliverables from uncritical ones and must be protected from delays in them by adding time buffers, which in turn make the plan robust against deviations from duration estimates. The approach supposedly leads to a reduction in rework and, therefore, of overall work and helps speed up a project quite dramatically. There are two caveats. Just like Agile methods, CCPM recommends dedicated teams, free from fragmentation and multitasking, that can focus on the project. I identified another caveat in classes. Developing a CCPM plan takes roughly twice the time compared that needed for a traditional PDM plan. CCPM promoters correctly say that this investment will be paid back when duration variation occurs. That generally leads to rewriting plans in PDM but is mostly captured in the time buffers of CCPM plans that can remain unchanged. The problems come when a plan needs alteration because a change request has been approved for the project, or because a plan was found unrealistic. The time and effort needed to change the plan increases.

Today, network diagramming is aided by computers with relatively inexpensive software, at least compared with the prices of the systems in use during the 1950s and 1960s, and even free solutions are available today (which have restrictions in functionality and comfort). They all simplify and accelerate the tasks of developing network diagrams, drawing them and doing calculations in them, tasks that are arduous when done manually. The free tools have the benefit of simplicity; you can learn them simply by using them. Advanced tools with their complexity help solve more challenging planning tasks and are faster to use in the hands of experts but need dedicated training[‡]. With all of these improvements from the past and the help of software, it is surprising that network diagramming is not used more frequently. In 2012, I tried another ad hoc survey on the question, "Are you using network diagramming?" in a LinkedIn group with project managers. At one point, I had to stop it. I had to accept its failure, because too many respondents had no idea what network diagramming was, and, therefore, did not understand the question, as I could see from the comments that they gave in addition to their answers. Their answers "polluted" the results and made them useless.

To make this clear: Network diagramming should be in the toolbox of every situational project manager to master projects and project situations that clearly require workflow definition

---

[*] (Sterman, 1992).

[†] (Goldratt, 1997).

[‡] I observe a shortage of good trainers in project management software, who understand both project management and the software and teach the second to support the first, not try to distort project management to make it suitable for the software.

and long-term prediction. It is a good practice and one of the most important tools, when "... it is necessary to take a strategic look at the entire job when planning, and to firm this strategy, in order to establish practical work sequences"*. My observation is that it is widely in use in construction, infrastructure, engineering, and to a lesser degree, in aerospace and defense. I found that it is not typically used in software development and implementation projects, organizational development, or marketing. Until some years ago, it was also not used in automotive product and production development. This seemed somewhat surprising, and I asked automotive project managers why they did not use network diagramming. The response was: "Mr. Lehmann, you seem to know nothing about our industry: We use different project management software than other industries, a product called RPlan, and this software does not support network diagramming". I met the principal founder of the vendor company of RPlan, a Munich-based company called Actano, years ago, and I asked him why the software does not support network diagramming. The answer was, "Why should it do that, network diagramming is not used anyway". Major changes have happened to the company since that time, and the software has been expanded to fully support network diagramming. Automotive development projects need a lot of prediction, and network diagramming has found its way into the offices of their project managers. It helps them perform better projects with better predictions and less rework.

Waterfall methods are generally implemented to meet stakeholder expectations; however, they sometimes serve as inexpensive excuses: "No, we will not implement the change that you desire. It is not in the plan, you know".

Some promoters of Waterfall methods also developed some kind of ideology around them, something I call "predictionism". They consider Waterfall approaches and methods generally superior, independent of the specific project situation, and insist on the implementation of aspects of predictive methods, such as long-term planning, top-down command-and-control structures, and rejection of stakeholder influence that may lead to change requests and to the need to re-plan major parts of the project. In their insistence on long-term predictability, they also develop an almost religious zeal (similar to the folks I referred to with Agilism). Again, they consider the world of project management a dichotomy with only two approaches: Waterfall and Agile. However, for these people the insistence on Waterfall methods comes with a fundamental rejection of Agile methods. Grey-shades between the extremes are not under discussion. If a person's only tool to earn a living is a screw driver, the person must convince people that the world is made of screws.

There is another aspect of predictive methods that is commonly overlooked: The plan must be complete, and the planners must not miss any work item during planning. An overlooked but necessary activity somewhere in the network, identified late, can make the entire schedule that has been derived from the network not feasible. A common example is rework after tests. Many network diagrams include testing activities and plans to immediately go on with their productive work after the test has been finished. They ignore the fact that the result of the test may be an identified need to fix errors and to do rework, for which no time has been planned and no resources have been booked. The assumption that all tests can be passed on the first try or that errors found can be fixed without impacting the schedule shows the planners may have been too confident.

---

* (Kelley and Walker, 1989).

4.5 Rolling Wave Approaches

Many projects are performed inside an organization's project portfolio. Small organizations may define all projects performed as just one portfolio, while larger ones may have more than one defined, each of them under control of another manager or management team. Portfolios can include internal projects, called for by an internal requestor and done for the organization's purposes or for an external customer who is paying for the project. A portfolio may also be a mixture of both types.

A company called Marten Inc. had a portfolio defined with three customer projects (for textbook simplicity—most portfolios are much larger), each with one deadline (also to keep the exercise simple). We call them Project A, Project B, and Project C and the deadlines also A, B, and C. The company made an assessment on where the projects stood and came to the following results (see Figure 4-4).

Project A and C are on schedule and able to meet their deadlines, but B is late. To make things worse, Project B is the project with the nearest deadline, and its project team is the one in trouble. Missing the deadline can lead to damage claims or contractual penalties*; it will make the customer angry and damage Marten's reputation. Meeting the deadline is both urgent and important. Marten made a decision to accelerate the project. De-scoping is not possible since Marten is working for a customer under contract, and other opportunities to increase the speed of the project have all been used to the maximum extent possible. The last option is to add more people, equipment, money, and so on to the project, something we commonly call "crashing". It is indeed possible for Marten to meet the deadline of Project B, and a sigh can be heard across the entire organization when the project is over. It was over budget, of course, consuming more employee working time than originally expected, but it was finished on time, and the customer could not reject the last payment.

---

* Depending on the jurisdiction and the contract.

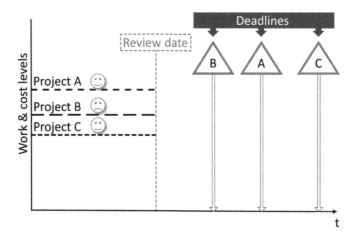

Figure 4-4   The project portfolio at Marten Inc. with three projects. Project A and Project C are on schedule, but Project B will miss its deadline.

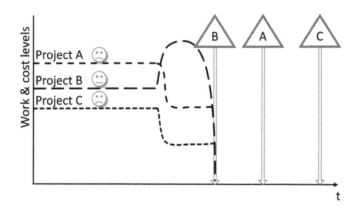

Figure 4-5   Project B met its deadline, but Project A and Project C are now in trouble and on the way to missing their deadlines.

Where did Marten find the additional resources? Marten's managers made a prioritization decision to focus on Project B, the crisis project, and took team members from Projects A and C, as shown in Figure 4-5.

The prioritization goes on: Project A is the next one that needs to meet a deadline, and resources are shifted there as well. When it is finished, Project C is crashed. Figure 4-6 shows the final situation of work and cost levels over the portfolios.

One may believe that with the end of Project C, the problems have been resolved. Marten may now be able to press a virtual reset button for its business and start being profitable. Unfortunately, the bids for Projects D and E have been won during that time, which helped the company secure future income but increased the already high stress level of those employees, who are active in projects.

The project managers finished their projects with a mixed perception of success and failure. They met the deadlines and avoided complaints for late deliveries, but the projects did not make the expected profits. For the entire organization, the overall results were even more detrimental for the following reasons:

- The "Law of Diminishing Returns" or "Law of Diminishing Marginal Utility" (also called Gossen's second law) says that when resources are added to existing resources, each of these new resources will add less value. A person thirsty in the desert may be prepared to pay a fortune for a glass of water. The fortune the person would be prepared to pay for the second glass would probably be much smaller, and so on. At one point in time, the person's thirst will be quenched, and the person will not need to pay anything for a glass of water. In the Marten example the productivity of a team will not grow proportionally with the increasing team size; instead, the team efficiency will go down. Members who are new to a team will take some time to become productive for various reasons, one being that they have to become familiar with their tasks, and while the incumbent team members have to train them, their productivity decreases as well.

- The "Law of the Minimum" (Liebig's law) says that once a bottleneck has been removed, there will be a new one. It may be office space, computer network bandwidth, shortage of tools, or another shortage of a resource that limits productivity and decreases the combined productivity of the team members.

- If a small number of people perform a large number of work items, it is easier to utilize these people. A greater number of people with the same number of work items will have more idle time while one person has to wait for someone else to have work completed that delivers input for the work of others on the team.

- Team members' motivation will suffer when they feel they are moved continually from one project to another one. The constancy of team situations that makes groups of people become "bands of brothers and sisters" vanishes; frustration and aggravation hold sway instead. Some will look for a more satisfying work place, and attrition of employees will worsen the situation, which is not sustainable.

While the crisis of one project is exported to another one in this portfolio domino effect, the efficiency of the entire organization suffers heavily. The result is a typical "Nash Equilibrium" (see Chapter 2), where the interests of the individual projects and the overall portfolio collide.

The case story shows how project managers have to expect changes from many dimensions. Continually changing requirements and new ideas from various stakeholders as to what they consider necessary and what is not required, volatile availability of resources, changing business situations of the organizations involved, and many more factors can make major parts of a project's predictions null and void after short periods of time. Project managers in these projects should be prepared. The changes may also come from inside the project. It may be impossible to perform some activities, and the project team has to find different solutions or workarounds. The team may have identified opportunities for simplification or improvement that they want to exploit for the project. The team may try to use benefit engineering (discussed in Chapter 5) to make budget or schedule overruns or other burdens acceptable for stakeholders by weighing them against additional benefits.

I mentioned John Fondahl, who was one of the original developers of network diagramming in the 1950s and 1960s. The focus of his Precedence Diagramming Method (PDM) was on manual application. The walls of "war rooms" in the following decades—when the method

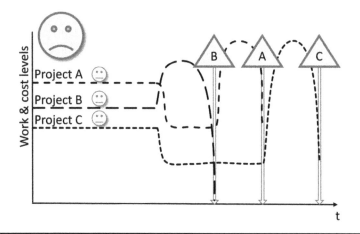

Figure 4-6   The final situation with three crashed projects.

found worldwide acceptance—were often filled with diagrams, including WBSs and network diagrams. Fondahl described a problem that emerged early with these plans: a lot of time and effort to develop, manually draw, and calculate were invested in the plans, and whenever a change occurred, major sections of the plan had to be re-developed, re-calculated, and re-drawn. The more energy and determination people put into making the plans originally, the more time and effort were needed when it became necessary to alter them. The effort and time needed for these changes grew with the length of the planning horizon and with the granularity and details covered in the plans, as both added to the size of the plan. One could ease the drawing work by using blackboards and chalk instead of room walls and India ink, which also made faster changes since the move replaced the razor blade used for scratching off the ink and a sponge to wipe the chalk drawings off. But this helped only partially. The fundamental problem remained that with growing predictiveness, the difficulty of keeping plans flexible grew. There are basically three solutions to the task:

1. **Use Computers.** In the early times of network diagramming, this was difficult. Computer time was expensive and so was developing the plans. The plans were not files used by a program, they were the programs, and each project plan needed to be programmed by a programmer. Data were mostly stored on punch cards, and a discipline needed to be developed to ensure that a deck of punch cards could not be damaged by negligence or accident; such a deck could lead to a crash of the computer, which was running several jobs at a time. If a crash occurred, the data would have to be recovered in a tedious and costly process. Today, software is much less expensive, and the risks have moved from damaged punch cards to untrained users. As in any discipline, mastership includes tool mastership, and project managers should be given the opportunity to learn their tools from competent instructors.

2. **Use "Rolling Wave" Approaches.** Picture a ship cruising across an ocean. Ahead of the ship is its bow wave, and while the ship is moving through the water, so is the bow wave some distance ahead of it. Rolling wave project management accepts the event horizons, the limitations of prediction and planning, without ignoring or rejecting the fundamental necessity of making projections. The event horizons of predictability may relate to time: the future is reliably predictable and can be planned only to a certain point in time, beyond which projecting would be turned into pure speculating. It may refer to granularity: The level of manageable detail in plans and forecasts is limited. The more detail one includes in plans, the more the plans will need to be based on assumptions that may later be proven wrong. It may be organizational because the project shares scarce resources with other projects inside a portfolio, depends on deliverables from other projects inside a program, or is a satellite project of a principal project (as discussed in Chapter 3), and the other project managers reject or neglect to inform others as to what is happening with their projects. An event horizon may relate to people, who have their own private agendas, desires, angsts, and other driving forces, to locations that are inaccessible, to technological advances, changing economic environments, and more. In the following discussion, I will focus on the time aspect, but the issue is much greater than just that. Effective rolling wave project management accepts these limitations and bases the decision on the planning horizons on two questions (see Figure 4-7):

   • 1st question: How much predictiveness does the project need?
   • 2nd question: How much prediction does the project allow?

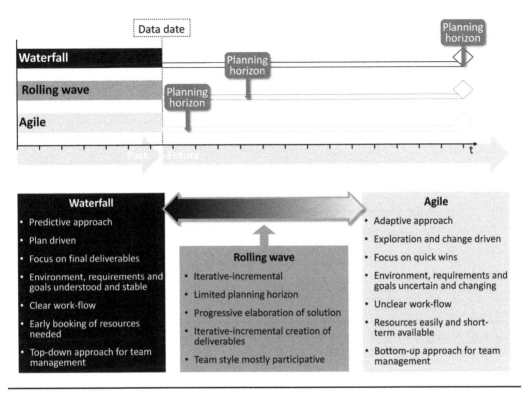

Figure 4-7   Rolling wave between Waterfall and Agile.

3. **Combine These Solutions.** Using modern project management software with rolling wave planning, it is possible to navigate a project between the two monsters of lack of predictability on one side and the inertness and debilitation that a project can have when a predictive approach has been taken too far, and when fear of having to change plans exceeds the desire to implement necessary or value-adding changes to the project. Applying rolling wave planning, the project manager reiterates the Plan-Do-Check-Act cycle along his or her own learning process and that of the team, adding increments of predictions when they are needed and possible. Software takes the burden from the project manager of having to do all the calculations and drawing work manually and also supports communication of revisions to the plan to the appropriate stakeholders.

There are, in essence, three forms of rolling wave project planning, as shown in Figure 4-8.

1. Rolling wave in strict, sequential phase-gate projects. In this approach, the project manager and the team take phase ends as planning horizons. This means that a current phase has been initiated and is planned until its end. Future phases may be roughly described but are not yet planned. It is quite common that at the end of a phase, a gate review looks at the previous phase to assess if all the work has been finished successfully by the right people, in the right order, at the right place, etc., and has been appropriately documented. This is the first part of the gate review, which focuses on formal phase exit criteria. If the reviewers are confident with the results of this part of the review, a

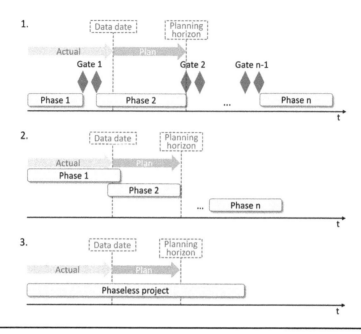

Figure 4-8   Rolling wave planning in a project with (1) strict, sequential phases with gates, (2) overlapping phases, and (3) no phases.

second part will be conducted, which focuses on the question, "Can the risks be taken to move to the next phase?" Only when the answer to this question is "yes" will the project manager get authorization to enter the next phase, which may then be initiated and planned. The benefit of this approach is the protection that gate keepers can provide. In pharmaceutical development, the gate keepers come from government agencies and protect the health and lives of people who participate in clinical trials by ensuring that only drugs that have been found effective and safe enough in the previous phase arrive at the next trial stage. In large capital projects, gates are often in place to protect the investors from monetary losses by asking if it is financially promising enough to continue the project and risk more money, or whether it would be better to stop the project at this point and accept the loss of money already invested. The price paid for the increased safety from strict phase-gate projects is that it slows down projects. Another price is that the project manager can rarely plan a future phase without the approval from the gate reviewers.

2. In overlapping phases, the risks are increased compared to the sequential phases, but the projects are completed more quickly if executed correctly. Rolling wave planning may still be linked with phase ends such in a way that currently performed phases are planned to their end, but future phases that have not yet been begun will be initiated and started after their start.

3. Many project managers feel restrained by phase models and run projects without phases. They will plan the horizon of rolling wave planning not according to a phase definition, but instead will ask which prediction horizon is possible and beneficial for the project.

These three types of rolling wave planning are not strictly separate. One may, for instance, use the third approach in a strict phase-gate model and plan for the next weeks in a phase that may last for several months.

Some advice on the use of software: Several of my customers have trained and dedicated experts for the software who act as assistants to the project managers. One such expert may be able to support three to seven project managers, depending on the complexity of the projects. Project managers in these environments do not waste their time trying to make the software do what they want, often a difficult task, and focus instead on their core tasks of coordinating, directing, and leading. The experts have a professional understanding of the software in use with its features, as well as with its limitations and weaknesses and a lot of experience in its application. Over time, they may learn several software products and are then able to select the software that best meets the needs of a specific project or has been mandated by a customer to better communicate with the contractor—and of course keep the outsourced work better under control. They also help PMOs develop unified approaches to project management across the organizational project portfolio that support project success, not impact it, as they are in continuous interaction with both the PMO and the project managers. The investment in these people pays back quickly to the employer by better cost-efficiency and reliability of the projects they support.

Prediction horizons may not be the same for all work items in a project. I mentioned earlier Wysocki's ACPF life cycle recommendation that moves from an Agile ideation phase to a much more predictive execution phase during the course of a project, which sounds like a reasonable approach for many projects. Another application of situational rolling wave could be to have different prediction horizons for different work streams. One stream, for example, may require highly skilled experts in a technical field that are hard to find and can be booked only for the long term in advance. Another work stream utilizes a workforce that can be obtained quickly and trained for the job. It may also be that there are several planning horizons defined for the entire project or its work streams. This approach implies a horizon for fine and detailed planning for the near future, and a high-level one that looks further into the future. On the move forward, the high-level ones will be refined, and sometimes may need to be changed, a process that is called "progressive elaboration" and reflects our learning curve along the project as shown in Figure 4-9.

A work stream is a collection of activities that together generate a set of connected deliverables. To give an example in an engineering project, one work stream may be to develop a

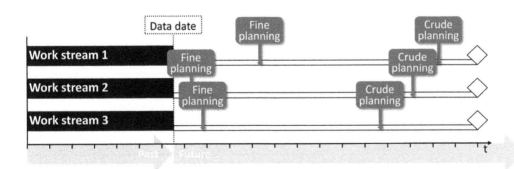

Figure 4-9   A project with three work streams and two planning horizons per stream.

piece of equipment, a second focuses on documentation and approvals, and a third may relate to developing environmental aspects such as making spares available in the warehouse.

How do event horizons of the project and planning horizons relate to one another? The first should be assessed by asking, "To what point in time can we effectively develop somewhat realistic and reliable plans?" One may ask the question repeatedly for different plans—for example, once for a fine plan and a second time for a high-level plan, and may ask the question also separately for the different work streams.

Two core elements of effective rolling wave planning are active refinement management and change request management. Table 4-1 shows the difference between the two forms of progressive elaboration.

A CCB is a Change Control Board, also known as Steering Committee, which is a meeting of key stakeholders to make decisions (not only) on changes that the team, the project manager, or the project sponsor cannot make alone. In a customer project, it may be advisable to have two CCBs: an internal one in which changes are discussed that are not the customer's business, and a joint CCB with the customer to discuss all change requests and other issues that need to be made together.

Change requests are two-faced, as they can rescue a project in a time of difficulty or even in a crisis, but poorly managed, they can also drive a project into crisis. For change request management or change control, one should have a process that has been formally agreed upon by the key stakeholders and that helps protect the project from harmful changes. Change

Table 4-1 The Distinction Between Refinement and Change

	Refinement	Change
Generation	The results of an iteration cycle during progressive elaboration	Requested by a stakeholder
Generally predictable?	Yes	No
First contact for request	Project manager	Preferably: Project sponsor
Changes requirements on scope, time, etc.?	Mostly: No	Mostly: Yes
Should impacts over knowledge areas be analyzed?	Often	Always
Requires a written change request	No	Preferably: Yes
Customer projects only: Should amendments to the contract be considered?	No	Yes
Should the change decision be escalated to the project sponsor or CCB?	Possibly	Possibly
Should the process be documented?	Yes	Yes
Can it turn a project from crisis to success and vice versa?	Unlikely	Yes

requests can have unintended side effects and secondary risks that should be assessed by subject matter experts before the decision is made to implement them or not. The last chapter has a recommendation on a protective change request process that can be implemented to shield the project from damages.

4.6 Connective Leadership and Achieving Styles

4.6.1 The Lipman-Blumen Achieving Styles Model[*]

Jean Lipman-Blumen[†] and Harold J. Leavitt published their Achieving Styles Model initially in 1983[‡]. Later, Lipman-Blumen linked this earlier behavioral work to her concept of "Connective Leadership" in her Pulitzer-Prize-nominated book, *The Connective Edge: Leading in an Interdependent World*[§], as part of a larger Connective Leadership™ Model. She described how the global environment in which leaders are acting has changed from the Physical Era, when leaders had to help followers satisfy basic needs such as hunger and protection, over a Geopolitical Era, when leadership consisted of emphasizing the differences from others, to the current Connective Era, which is driven by global alliances and networks. The Connective Era poses challenges for leaders that are different from those in past eras. Connective leaders ably navigate their way through a continuous tension field in which diverse individuals, groups, and organizations must continuously work and live interpedently. The Achieving Styles Model is supported by a vast amount of empirical data from global leaders, and it is based on a simple question: What shifting combination of behaviors do leaders need to use to achieve their goals? The next steps involve measuring these behaviors or "achieving styles", building profiles from the results of these measurements, and connecting these profiles with leadership situations and their specific needs.

It is important to understand that the model is not based on personality—in contrast to many other models that are similar at first glance—but on behavior. The question under discussion is not who people are, but what they do, and, especially, what they like and dislike doing. Leaders like certain styles because they find themselves applying these styles successfully, and they dislike others because they believe they would probably fail if they used them.

Behaviors are open to change, even to self-invention, while personality is more static. Behaviors can be learned and unlearned in a relatively short time, while personality changes only over long cycles and may not change. Behaviors are of course influenced by personality, but there are many other factors that influence them, such as experiences, feedback from other

---

[*] "Achieving Styles" and "Connective Leadership" are trademarks of the Connective Leadership Institute, Pasadena, CA, USA.

[†] Dr. Jean Lipman-Blumen is the Thornton F. Bradshaw Professor of Public Policy and Professor of Organizational Behavior at the Drucker School, Claremont Graduate University in Claremont, CA, USA. She served as assistant director of the National Institute of Education and as special advisor to the Domestic Policy Staff in the White House under President Jimmy Carter. She founded the Connective Leadership Institute in Pasadena, CA, USA, with Professor Harold J. Leavitt, Kilpatrick Professor of Organizational Behavior, Stanford University, Graduate School of Business.

[‡] (Lipman-Blumen, Handley-Isaksen, and Leavitt, 1983).

[§] (Lipman-Blumen, 1996).

people, cultural norms, and even laws. Drugs and mental disorders may influence behaviors, as one may remember in Robert Louis Stevenson's novel *Strange Case of Dr. Jekyll and Mr. Hyde.* Sometimes, leaders change their styles unintentionally in response to particular situations. I remember a leader, who had a strong entrusting style, scoring around six measured on a scale from one to seven, but got heavily disappointed by an important business partner, and the person's score on the entrusted dimension fell sharply. After four weeks, the score was back to the person's normal value.

Changing an individual's personality cannot be the business of a project management trainer as I am. Helping project managers to acquire and develop behaviors that would be beneficial for them and their stakeholders, and supporting them in applying these behaviors with confidence and joy in the right situations, definitely is.

Lipman-Blumen and Leavitt identified nine Achieving Styles, which she classified in three groups, as shown in Figure 4-10.

- **Direct** styles are applied by leaders whose behavior is mostly characterized by individualism and expressed through behaviors that emphasize confronting challenges on one's own, mastering one's own tasks, competing, and taking control and organizing.

- **Instrumental** achievers accomplish their tasks by using themselves and others as instruments. They do this in three general ways: One, they use their own personal qualities

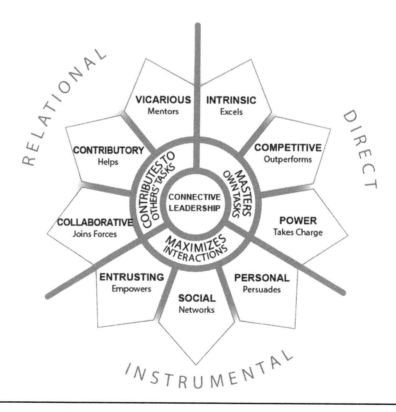

Figure 4-10   The nine Achieving Styles clustered in three groups. (Diagram provided by and used with permission of Jean Lipman-Blumen, ©1996.)

and background to attract others to help them in their efforts (personal). Two, they select individuals with appropriate background, knowledge, experience, skill, or contacts to help in task achievement (social). Three, they rely upon any of the individuals in their known environment, with or without specific relevant expertise or contacts, to carry out the work (entrusting).

- **Relational** styles are implemented through trustful rapport and are based on teamwork, contributing behind the scenes, and mentoring or simply deriving a sense of achievement through others' accomplishments. When facilitating or mentoring, leaders often stand back behind others whose success they support.

No leader uses only one style, but one often finds a style that a person prefers to use and champions proudly. Leaders also may behave differently in distinctive settings; for example, a highly competitive manager or politician may be the most helpful father or mother toward his or her children.

These are the nine Achieving Styles, as discussed in the following sections.

Direct Styles

- **Intrinsic style.** Intrinsic Direct leaders prefer to master tasks by themselves. When they delegate a job to another person, they often have the impression that they could do the work much better than their subordinate. They do not compare themselves to others and are their own harshest critics. In achieving situations, they do not care much about what others think of their accomplishments, but, instead, measure their performance by their own metrics.

- **Competitive style.** Competitive Direct leaders try to outdo others. In their minds, they hold scoreboards and rankings, and it is important for them to be on the winning side and rank at the top. They prefer achieving situations in which players compare their forces directly and have opportunities to knock out opponents. Competitive leaders measure their performance by their ability to excel beyond and even disrupt actual or perceived rivals in their attempts to achieve, and by how much they could benefit from their competitors' weaknesses.

- **Power style.** Power Direct leaders like to take charge of everyone and everything. They like to bring order out of chaos. They prefer being in authoritative positions, where the link between their investments and the achieved returns are easily seen. They want people under their domain simply to follow commands and function as expected, and they describe what they perceive as underperformance under their command in clear words. They prefer achieving situations with a clear team hierarchy and in which people involved have to follow well-established procedures and chains of command. Power Direct leaders measure their own performance and success along return-on-investment calculations for tangible and intangible achievements.

Instrumental Styles

- **Personal style.** Personal Instrumental leaders are perceived as shiny and magnetic by their followers. They build an aura of charisma around them, often using their families, ancestors, educational and occupational achievements, or distinctive events in their lives that make them special. They use humor and convincing charm to make others support their case, and when they make jokes, they are more likely to engage in self-deprecating humor than to have fun at someone else's expense. They often prefer achieving situations in which an individual can support a team, but can also take an exposed position from time to time. Personal Instrumental leaders consistently measure their degree of achievement by the intensity of admiration and confirmation that they get from others and the degree to which others would like to associate and work with them.

- **Social style.** Social Instrumental leaders delegate the right tasks to the right people. They know a lot about the skills, experience, networks, and interests that others have and consider them when they assign people to open jobs. They show interest in people who, in turn, strive to perform as expected. They also connect with each other to strengthen the networks that help them achieve. If they excel, it benefits others in their networks as well. They measure their degree of achievement by the functioning of their networks.

- **Entrusting style.** Entrusting Instrumental leaders prefer to delegate as well, but they often send people on stretch assignments that pose requirements beyond these peoples' qualifications and experience. Then, they expect people to grow and learn with the tasks. They often allow the individual assigned to the job to make an error, but not to repeat the error. Their entrusting style puts people under pressure, who then must give something back so as to not disappoint the leader: results. Entrusting Instrumental leaders often perceive disappointments from trusted people as especially painful. Entrusting Instrumental leaders measure their degree of achievement by the delta of capabilities that people under their domain have achieved to meet the expectations set for them.

Relational Styles

- **Collaborative style.** Collaborative Relational leaders prefer doing the job together with people with whom they have developed trust and rapport, possibly through long-term work relationships or teamwork. They like working in groups, to whose members they will offer loyalty and dedication, and they will place the group's goals above their own. When a new task must be done, they wonder who would be the best individual(s) with whom to join forces. In turn, they expect to receive their share of the reward. Collaborative Relational leaders prefer achievement situations, in which the team is more important than the individual, and where the success of the team depends equally on the contribution of all team members. Collaborative Relational leaders measure the degree of achievement by the extent to which goals have been met jointly by the group to which they have contributed. They also accept their share of responsibility and blame for failure.

- **Contributory style.** Contributory Relational style leaders prefer to come second and act in the background. They take pleasure in the successes of others as long as the successes can be traced back to their active support. When others need help, they take joy in providing this help. In sports, they prefer to assist others and help them to score rather than scoring themselves. It also may be that they are more active in developing a new sport by setting the rules and providing the infrastructure than actually playing the sport. They measure their degree of achievement by their active involvement in the successes of others.

- **Vicarious style.** Vicarious Relational leaders also take joy in the successes of others, but of people in whose success they had no part. As mental rather than physical supporters, they give others motivation, inspiration, and the feeling that someone cares for them and acclaims their achievements. They do not expect to be given credit when others achieve; the sense of pride that they take from the achieving situation is sufficient. In achieving situations, they may prefer to be observers, rather than players, and when the teams that they support achieve, they feel as if they have achieved. They measure their degree of achievement by the achievements of those with whom they identify and support.

4.6.2 Application of the Lipman-Blumen Achieving Styles

Lipman-Blumen says that leaders generally perform best if they are confident and able to master all nine Achieving Styles. In that way, they have the ability to adjust to the needs of a variety of challenges. This was confirmed by Verna Wangler in her dissertation[*], which showed how leaders with complex behavior are more likely to be effective, as they can master a wider range of situations by adapting behaviors to what is situationally appropriate. This is probably true for any manager. It is especially interesting in project management, however, where during the lifecycle of the project, its internal dynamics meet the external forces of the performing organization and the multi-faceted environment in ever new combinations. As discussed in Chapter 1, project management is an open-skill, not a closed-skill, discipline, and during the interminable confrontation with the often unexpected but ever-changing realities in and around the project, effective project managers are separated from their less effective colleagues.

The Connective Leadership Institute, in Pasadena, California, uses a suite of eight leadership assessments to generate individual and organizational connective leadership profiles. Surveys are constructed on a 7-point Likert scale, ranging from "never" to "always" uses the style. The profiles are shown in the form of polar graphs that provide insight into the behavioral strengths and weaknesses of a person, as Figure 4-11 illustrates. It shows the results of an Achieving Styles Inventory (ASI), an instrument that is used to measure a person's individual Connective Leadership profile.

These are the eight Connective Leadership/Achieving Style Assessments[†]:

1) **ASI.** The Achieving Styles Inventory measures the individual's Connective Leadership profile and gives the person a better understanding of its current behavioral setting.
2) **A-ASI.** The Aspirational Achieving Styles Inventory measures the individual's Connective Leadership profile that the person aspires to have in the future.
3) **OASI.** The Organizational Achieving Styles Inventory measures the leadership behaviors that an organization values.
4) **A-OASI.** The Aspirational Organizational Achieving Styles Inventory measures the leadership behaviors that individuals would like to see valued by the organization in the future.

---

[*] (Wangler, 2009).
[†] (Connective Leadership Institute, 2015).

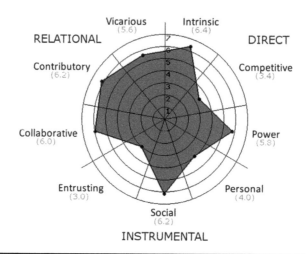

Figure 4-11   An ASI profile of a person who is mostly confident using the Intrinsic, Social, Contributory, and Collaborative styles, but less confident using Entrusting, Competitive, and Personal styles. (*Source:* Connective Leadership Institute, published with its permission.)

5, 6) **ASSET.** The Achieving Styles Situational Evaluation Technique (ASSET) measures the requirements ("demand scores") of a role or project in terms of the Achieving Styles required for that specific role or project. There are two versions of ASSET, one for a role called ASSET (r) and one for a project called ASSET (p).

7) **ASI360.** A 360-degree assessment of the Leadership Styles of a focal person[*]; the instrument measures the Connective Leadership profile by comparing an individual's self-assessment (ASI360F) of the focal person's behavior made by relevant people from that individual's environment (ASI360E).

8) **A-ASI360.** The aspirational 360-degree assessment—what future Connective Leadership profile does the focal person aspire to the future (A-A360F), and what profile would relevant people from the individual's environment aspire to that person (A-ASI360E)?

The assessments may be displayed as overlays of two graphs (see Figure 4-12) to show deltas between results. These overlays offer new and valuable insights into the behavioral matters of actual, intended and needed leadership behaviors. They help identify needs for development on a personal as well as on an organizational level and support the more situational allocation of people to tasks.

4.6.3 Real-Life Examples and Application in Project Management

Figure 4-12 shows two examples of how Connective Leadership profiles can help understand leadership behavior and help leaders master leadership situations.

---

[*] A focal person is in the focus of the assessment, and the person's self assessment is compared with assessments by the environment.

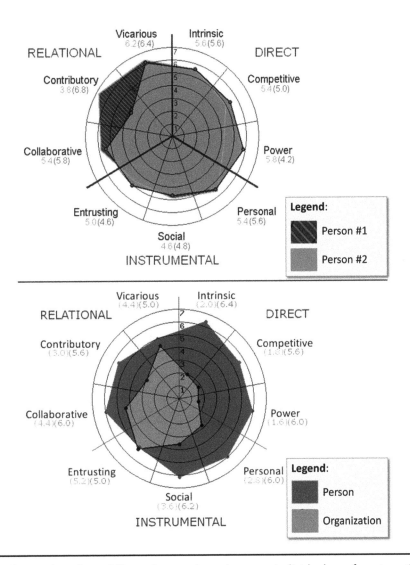

Figure 4-12 The overlay of two ASIs on the top shows how two individuals perform together as a team. On the bottom is an overlay which shows the ASI of a person and an OASI of the employer. (*Source:* Connective Leadership Institute, published with its permission.)

The top example shows two leaders from project management, who have worked together as a successful team over many years. The overlay of the two individuals shows some interesting peculiarities:

- In seven out of nine Achieving Styles, the two leaders are very similar. These are leadership behaviors in which they experience harmony and empathy for each other. They act similarly in leadership situations that require these behaviors.

- Too much harmony would make people inert. Contributory and Power styles prevent the two from becoming sluggish. The two have some tension here that keep them alert and force them to explain the reasoning for behaviors to each other.

- One can also see that the two are probably more effective as a team if Person #2 stands in front, where the person can apply the strong Power style, and when Person #1 steps back behind Person #1 and applies the strong Contributory style.

The bottom example was generated from an individual and an organizational assessment done by a person whom I did not know at the time. A friend of mine called me and asked if I would be prepared to do the assessment with a person from his environment anonymously, and we did it. This overlay is the opposite of the previous one:

- In seven out of nine Achieving Styles, the person feels that the efforts are not rewarded adequately by the organization.

- Only in two Achieving Styles is there harmony between the effort invested by the person and the tangible and intangible rewards by the organization.

My impression: Effort-Reward Imbalance (ERI) is a classical cause for distress and team attrition. My assessment: The person was about to leave the organization. When my friend called me again, I was careful however, as I did not know who it was. It could be his secretary, and being too open, I could cause more damage than benefit. I told him that the profiles show "a situation with severe conflicts". He responded: "Yes, it is my wife, and she has a lot of problems in her job. She is going to quit soon". The focus person was a key woman in the IT organization of her employer, and it would probably have been interesting information for the employer to know how unhappy she was in her job and why.

The "well-rounded" project manager related to the Achieving Styles, of course, a person who can adjust to any leadership situation, would be a desirable leader for any project, but such people are not common; moreover, Achieving Styles can change in response to certain situations, as described before. How should projects then be staffed to be prepared for a variety of leadership challenges? Figure 4-12 gives an answer to this question as well: Combine people with diverse strengths and weaknesses. While the Connective Leadership profiles are great tools to identify the best matches, a lot can be done with the basic understanding and the application of common sense. Finally, the Connective Leadership/Achieving Styles Model is based on rigorous scientific validations and behavioral studies conducted over a 10-year period. It describes behaviors that we can observe around us, and with a well-formulated model based on empirical research, it is much easier to understand and accurately predict the behavior of our stakeholders.

4.7 Favorable and Detrimental Practices

I mentioned earlier my research project, which I did with a group of 17 seasoned experts in project management. The first part was to develop a typology of projects. In essence, the Lifecycle Approaches and Achieving Styles described in this chapter are also typologies of behavioral dimensions. The second part of the research therefore dealt with the question, "Are there are favorable and detrimental combinations of project types with these behaviors?"

The experts used a system with points ranging from –3 to +3 that led to evaluations as was shown in Figure 4-13; the following diagrams show the average results of their assessments.

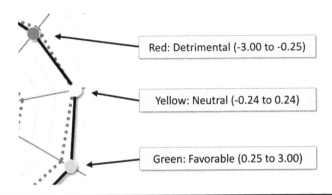

Figure 4-13   The classification of the results follows a ±3-point system between detrimental and favorable.

4.7.1 How Can the Following Information Be Used Best?

I first recommend using these data as guidelines. They are based on centuries of cumulative experience by seasoned project management experts, but there are more factors in a project that influence the dynamics of success and failure than the nine dimensions of project types that the experts had identified. I further recommend considering this information rather as warnings than as recommendations for action. Detrimental behavior is quite certain to lead to failure. In comparison, favorable behavior can much less be a guarantee for success. The don'ts are probably more important in the list than the dos. Avoiding major failures opens opportunities for success, but it is not yet true success. However, failures close doors for success.

If you have a project that you wish to assess, it is recommended that you go through the following pages and ask yourself, for each of the types listed, if your project belongs to it. Is it a Mark 1 or Mark *n* project, a Greenfield or Brownfield project, and so on. Then make a note for every yellow spot and especially for every red spot that you find. At the end, you have a list of neutral practices, whose application are likely to be a waste of time for the project, and detrimental practices, that can drive the project into trouble.

The graphical representations below are a quick and practical help for practitioners in project management. If you are interested in the details of the research, including the detailed process and the numbers behind the graphics, please contact me. I will be happy to give you access to the much more detailed original research paper.

A conclusion of the project types and how the practices perform with them is that the combination of Rolling Wave approach and a focus on the instrumental and relational achieving styles seems never to be wrong. This does not mean that Powerful and Competitive Achieving Styles are never necessary—we surely have to use them with situational care.

An interesting point about the Rolling Wave approach is that its backers can easily shift their behaviors toward a more Agile and adaptive or a more predictive approach, depending on the situation. It is very hard for Agile people to manage a project that needs and allows for high predictions, and it is also difficult for the champions of Waterfall approaches to change to more Agile methods. Both parties often complain about the gap between them, unaware that Rolling Wave approaches have already built bridges over the gap—decades ago, by the way. Rolling Wave projects can be easily moved to a more Agile side, when the predictability and the needs for predictions are low and move the planning horizon further into the future, if this is needed and possible, and turn the project into a Waterfall project. Among the life cycle approaches in the research, Rolling Wave was not detrimental for any type of project. However, it is not a Best Practice but is a great starting point, which is suitable for probably more than 50% of the projects (see Chapter 3), and it is also a basis from which one can easily turn to more Agile or predictive approaches and methods, depending on the situation. I will go into more detail about the practical application of this concept in Chapter 5.

Mark 1 Projects and Mark *n* Projects

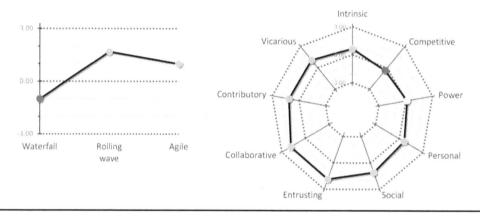

Figure 4-14   For the novelty in Mark 1 projects, predictive Waterfall approaches and competitive Leadership Style are detrimental. Power style is neutral. All other approaches and Achieving Styles are favorable.

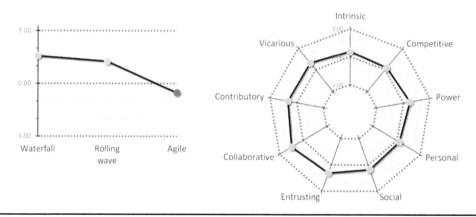

Figure 4-15   Mark *n* projects can be derived from former projects. Agile approaches are detrimental. All other approaches and Achieving Styles are favorable.

Greenfield Projects and Brownfield Projects

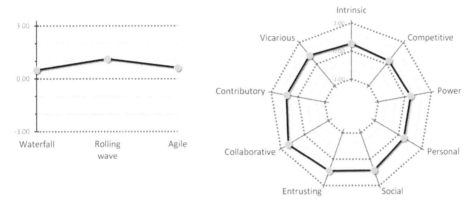

Figure 4-16   For Greenfield projects, all approaches and Achieving Styles are favorable.

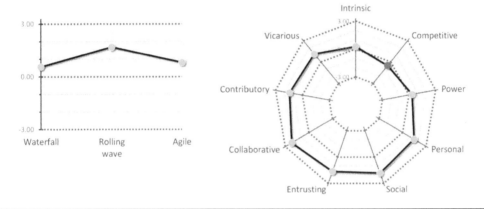

Figure 4-17   Brownfield projects have to consider many stakeholders. Competitive Achieving Style is detrimental. All other approaches and Achieving Styles are favorable.

Siloed Projects and Solid Projects

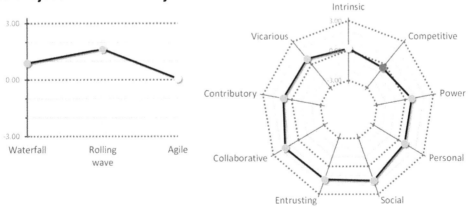

Figure 4-18   Siloed projects are fragmented. Competitive Achieving Style is detrimental, Agile methods and Intrinsic Achieving Style are neutral. All other approaches and Achieving Styles are favorable.

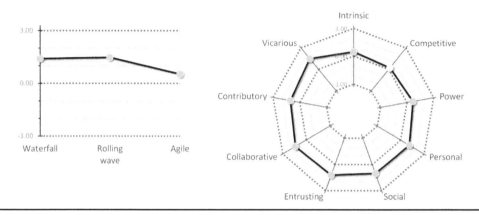

Figure 4-19 Solid projects have low fragmentation. They are neutral to the Competitive Achieving Style. All other approaches and Achieving Styles are favorable.

Blurred Projects and Focused Projects

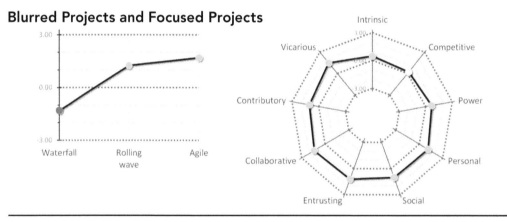

Figure 4-20 Blurred projects are not clearly defined and separated from other tasks of the organization. Waterfall is detrimental for them. They are neutral to the Competitive Achieving Style. All other approaches and Achieving Styles are favorable.

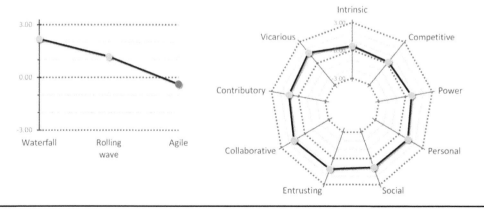

Figure 4-21 Focused projects have clear boundaries and definitions. Agile approaches are detrimental. All other approaches and Achieving Styles are favorable.

High-Impact Projects and Low-Impact Projects

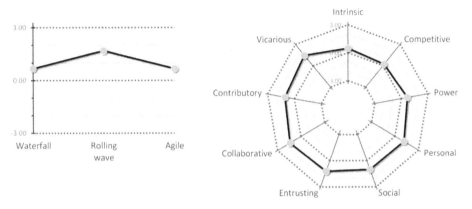

Figure 4-22   For High-impact projects, all approaches and Achieving Styles are favorable.

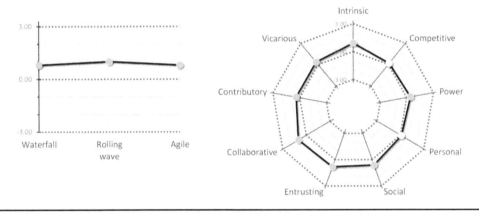

Figure 4-23   For Low-impact projects, Competitive and Personal Achieving Styles are neutral. All other approaches and Achieving Styles are favorable.

Customer Projects and Internal Projects

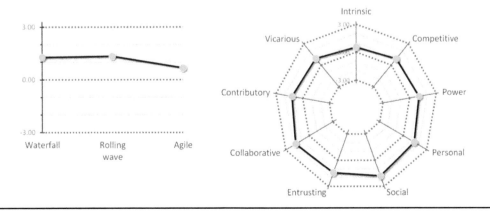

Figure 4-24   For Customer projects, all approaches and Achieving Styles are favorable.

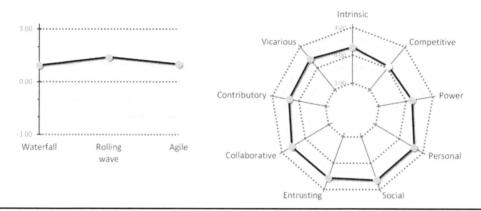

Figure 4-25   For Internal projects, Competitive Achieving Style is neutral. All other approaches and Achieving Styles are favorable.

Stand-Alone Projects and Satellite Projects

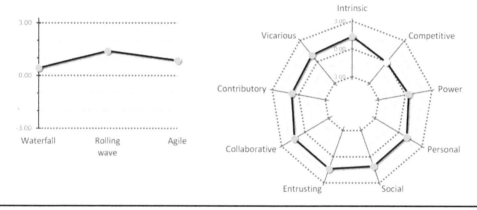

Figure 4-26   For stand-alone projects, Competitive Achieving Style is neutral. All other approaches and Achieving Styles are favorable.

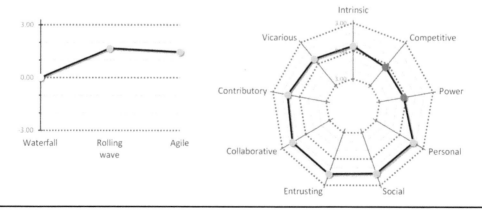

Figure 4-27   Satellite projects are performed in the wake of principal projects. Competitive and Power Achieving Styles are detrimental; Waterfall methods are neutral. All other approaches and Achieving Styles are favorable.

Predictable Projects and Exploratory Projects

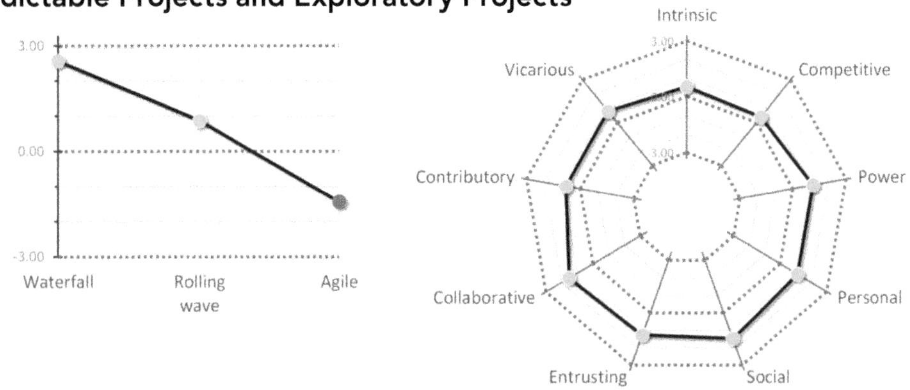

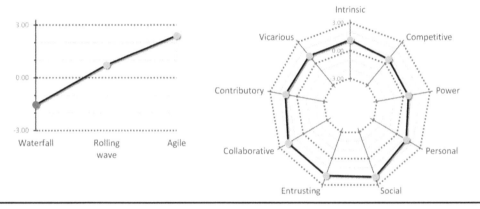

Figure 4-28   Predictable projects allow long-term predictions and plans. Agile approaches are detrimental. All other approaches and Achieving Styles are favorable.

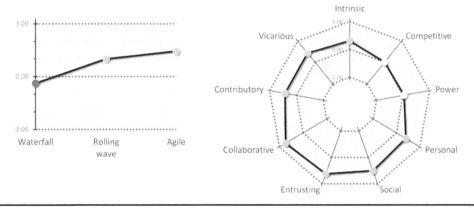

Figure 4-29   The way is made by walking in Exploratory projects. The predictive Waterfall approach is detrimental. All other approaches and Achieving Styles are favorable.

Composed Projects and Decomposed Projects

Figure 4-30   Composed projects are developed bottom-up by their contributors. Waterfall is detrimental for them. They are neutral to the Competitive and Power Achieving Styles. All other approaches and Achieving Styles are favorable.

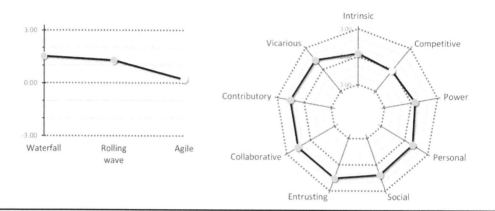

Figure 4-31 Decomposed projects are developed top-down and work is then assigned to contributors. Agile is neutral for them. They are also neutral to the Competitive Achieving Style. All other approaches and Achieving Styles are favorable.

Chapter 5

Some Basic Tools for SitPM

There is a wide body of literature on methodical and software tools in project management, and it is not the intention of this book to repeat what has already been written in more than sufficient detail by others. The focus of this chapter is on tools that are especially valuable for situational project analysis and management, possibly with some expansions or alterations to already existing descriptions.

5.1 Introductory Questions

Again, some of the following introductory questions relate to the tools that will be discussed in this chapter and may be difficult to answer before you have completed it. While some tools may be known from other sources, there are some specifics in the descriptions here that are different and make the tools appropriate for situational approach to project management.

You may try the questions right now, or read the chapter and then answer them.

---

1. Which of the following is not a purpose for having an assumptions register?

   a) An assumptions register can help easily identify project risks.
   b) An assumptions register can speed up the team-building processes.
   c) An assumptions register can save the project team from factoids.
   d) An assumptions register can provide a common understanding of why decisions are made.

---

2. Which approach may help a project manager and the team in a customer project deal with project difficulties from unrealistic cost estimates, time commitments, unrealistic technical solutions, and other contractual obligations that the project is unable to meet?

   a) De-scoping
   b) Crashing
   c) Cost engineering
   d) Benefit engineering

---

3. A high span width in Stakeholder Force-Field Analysis (StaFFA) indicates for the project that:

   a) the project manager and the team will probably lose the support of senior management during its lifecycle

   b) the project will be easily finished as planned and will meet all cost, time, and result objectives that have been agreed upon by key stakeholders.

   c) during the project, the project manager and the team will probably stand in the middle of a massive conflict.

   d) the deadlines of the project are difficult to meet due to a high degree of resistance by key stakeholders.

4. Which of the following are true for change requests?

   a) Change requests have the power to get a project out of trouble, but also to drive it deeply into crisis.

   b) Change requests are a sign of poor planning and should generally be avoided by the project manager and the team.

   c) All change requests should be welcomed by the team, as they provide the opportunity to please the client.

   d) If a change order comes from the sponsor, the impacts should not be assessed and the order immediately executed.

5. Which of the following are true for estimates under pressure?

   a) Having to develop estimates under pressure separates the men or women from the boys or girls.

   b) Estimates under pressure are often political, and people avoid doing them as they are often blamed if the prove to be incorrect.

   c) Estimates made under pressure are generally overly optimistic.

   d) Estimates made under pressure are generally overly pessimistic.

6. You are planning a kick-off meeting. What should you take care of now?

   a) Plan a brief and informal meeting, which may be repeated easily if necessary.

   b) Take the opportunity to review and update project plan documents before the meeting.

   c) Invite attendees short-term to make sure that you are the best prepared person.

   d) Keep the agenda open so that all objections may be raised and discussed during the meeting.

5.2 Stakeholder Force-Field Analysis (StaFFA)

Force-Field Analysis was developed by the German-American psychologist Kurt Lewin and first published in 1943[*]. It is simple in its application, but powerful in its ability to predict if a change will be difficult or easy to manage. In the application shown here, it also gives the

---

[*] (Lewin & Weiss, 1997).

project manager and the team an indication if they have to expect a lot of conflicts or will perform the project in a more harmonious environment, and it provides some indication on how to adjust approaches.

The model is built on a simple metaphor: Some people want to pull an item in one direction, others want to keep it in its place. Which group will be stronger? Both groups will influence the ability of the project to bring about the deliverables and changes for which the project is undertaken, and the balance between them can be the most important factor in the dynamics of success and failure.

There is a slight difference in the approach that I recommend in what I call StaFFA to what Kurt Lewin recommended. He includes environmental factors in his analysis that accelerated or slowed down change, something one may also do for a project. My focus, however, is on people. I am also not only interested in the "Force Field" that he described, but also in the tensions that the analysis visualizes in the span between strong drivers and restrainers.
The process steps are as follows:

1. Build a work group, commonly between three and seven persons; five is a good average size. Form a multidisciplinary and collocated focus group for the analysis.

2. Close the door of the meeting room. Then ask the group to list stakeholders, such as people, groups, or organizations, who may drive the project or restrain it. Write the names on the side of a page (flip-chart or whiteboard are fine, or use a PC with projector).

3. Assign force numbers to the stakeholders between –10 (strongly restraining) and 10 (strongly driving). Then draw a diagram from this information as shown in Figure 5-1 to represent the numbers graphically.

Driving forces in StaFFA are those stakeholders who generate process in the project. This is mostly done through their active contribution and their preparedness to act as a part of a greater system. Restraining forces slow down the project and may bring it to a halt if they are

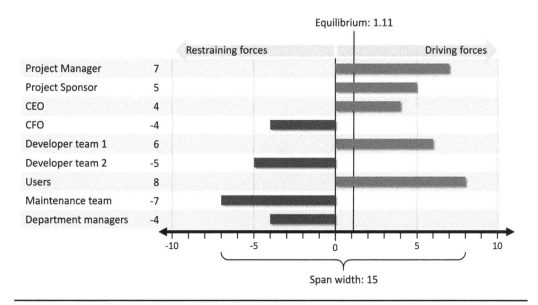

Figure 5-1   A StaFFA with nine stakeholders.

stronger than the driving forces. Restraining forces may be stakeholders who resist the project actively or passively by rejecting the project and not providing the degree of support that it needs. Team members or contractors who basically support the project but are unreliable or incompetent in performing their job or whose availability is uncertain from factors beyond their control and that of the project manager may also restrain the project. People who bring conflicts into the team and undermine the morale of others are another example.

Two values are interesting for the project:

- **Equilibrium.** The arithmetic average of the numbers assigned to the stakeholders is a metric for the likeliness of the project to succeed. In the example, the number 1.11 is positive, so the driving forces are stronger than the restraining forces. The team should be able to perform the project successfully. The number is not very high, so the team needs to take its time to surmount the restraining forces, which means it will probably not become a superfast project.
- **Span width.** Fifteen steps, (from −7 to 8) on a scale with 20 steps is quite a lot. The project will probably have a number of conflicts, and the project manager and the team will be in the center of the conflict.

Conclusions that one can draw from the results of stakeholder force-field analysis:

- **Negative equilibrium.** The restraining forces are stronger than the driving one. The project is likely to fail.
- **Low positive equilibrium.** The driving forces are stronger than the restraining forces, but not much. The project schedule should allow time needed to manage resistance.
- **High positive equilibrium.** The project will probably be a quick one.
- **High span width.** The project is bound for conflicts. Clear agreements at the onset of the project to have clear authorizations (Project Charter) and definition of project management processes and responsibilities (Management Plans) will be important later to master these conflicts. Constant awareness of conflict potentials among stakeholders and attention to disagreements and tensions among them is also advisable.
- **Low span width.** Just do the project and rely on the consensus and common mission of shareholders.

One should repeat the analysis from time to time to reflect the changes in the force field and assess the results of measures taken by the team to reduce resistance and other forms of restraining forces.

5.3 Benefit Engineering

Many projects are successful. Others are initiated and sent on a course bound for inevitable failure right from the start. Some are Zombie projects that had no chance for success at any time, and it is often impossible to salvage them later. The problems may be technological, organizational, or interpersonal in nature or simply that sales people have sold something

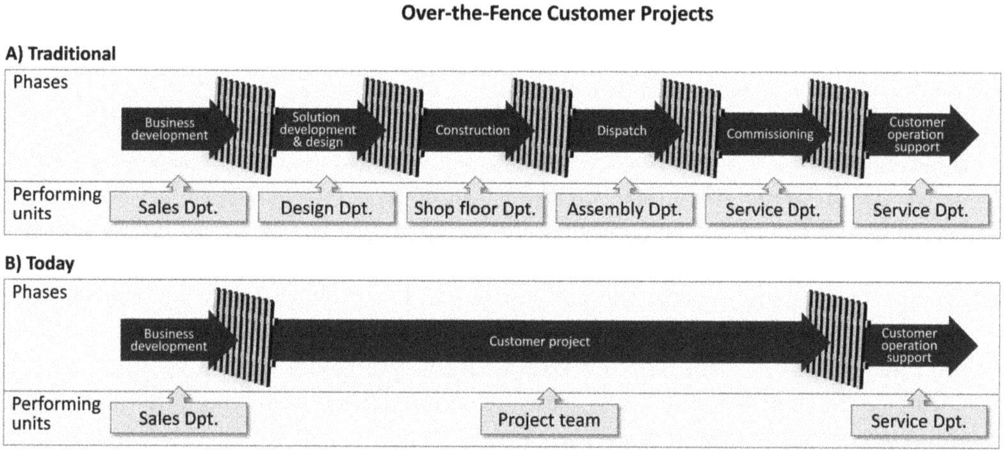

Over-the-Fence Customer Projects

A) Traditional

| Phases | | | | | | |

Business development → Solution development & design → Construction → Dispatch → Commissioning → Customer operation support

| Performing units | Sales Dpt. | Design Dpt. | Shop floor Dpt. | Assembly Dpt. | Service Dpt. | Service Dpt. |

B) Today

| Phases | | |

Business development → Customer project → Customer operation support

| Performing units | Sales Dpt. | Project team | Service Dpt. |

Figure 5-2   An example for an over-the fence project from engineering, as would have been commonly performed before project management was introduced and today. There are often still two fences left.

such as a flying carpet to a customer. Others are not fundamentally wrong, but the data do not fit. For example, a customer project bid or proposal may have been won based on extreme lowballing and on schedule promises that were unrealistic right from the start. A frequent cause for problems in managing customer projects is information that has not been communicated to the project manager in its entirety or has already been so vast at the onset of the project that the project manager and the team were unable to grasp it all in the short time available. Often the pressure to quickly become productive is too high, colliding with the time needed by the project manager to become familiar with the project, its environment, and its stakeholders. Earlier, projects were often conducted by business units such as departments or branches "over the fence" (see Figure 5-2A), and according to literature, this time should be over, but in the reality of customer projects, two fences often have survived the change inside the lifecycle of the business with the customer, at the beginning and at the end of the project, as is shown in Figure 5-2B.

The fence between the sales department and the project team often has the consequence that projects during business development or during project selection are set up with expectations, promises, and obligations that cannot be met. The price in fixed-price contracts may be found later to not cover costs, deadlines may be impossible to meet, technological demands may be overly challenging, and the promise to disrupt operations by no more than four weeks may be found to be too optimistic. Optimism is part of the profession of a good sales person. The ability to credibly communicate the "Easy" proposition to both the customer ("We are so easy as a vendor and will save you from all problems. Select us!") and the performing organization ("This prospect is so easy, things will not go wrong. Remove the contingency premium from the offered price!") is an important skill, which helps keep the company in business. However, it can later become a problem when the reality is not easy at all. It will then be the responsibility of the project team to find a solution.

Similar problems occur in internal projects. Here, it is rarely due to sales people but to the people involved in developing requests and business cases for projects and those who decide upon them. It is rather rare that project managers will get the time, funding, resources, and other resources they need, and they face the additional problem that most projects here are in a weak matrix, which will make it difficult to obtain resources as needed and give the operational managers a strong stance in their defense against disrupting daily operations, which may be disturbing but necessary. Your project is on the way to foreseeable crisis, and you will be the project manager in charge when this happens. Running away is not always a bad decision, but at some point, it is simply too late, because all the blame for the bad project will then be on you, and your reputation, self-respect, and most important, your professional resume will suffer from this negative assignment. You will notice that in search for a new job, recruiters will be more reluctant to invite you to an interview, and the interviews will be much more difficult for you. Sometimes, it is surprising how badly informed recruiters can be, but then we meet others who are perfectly informed, and these are commonly the ones who have the hands on the most attractive project assignments.

For both customer and internal projects, a common approach in situations in which projects overrun budgets and deadlines is cost engineering. The project team tries to reduce the costs, effort, and time needed by reducing the scope or shifting some parts of the scope back into the responsibility of the customer or the internal requester. Work may be done with reduced meticulousness, assuming that the first 80% of the results will take only 20% of the time, effort, and costs, and who notices the other 20%? Expensive components may be replaced by less expensive ones, and the organizational overhead is reduced to focus more on the productive resources. Figure 5-3 shows the focus of the cost-engineering approach.

There are limitations for cost engineering: In a customer project, the contractor, probably your employer, does not want to step into a breach of contract situation. Most organizations do not want to run into a breach of contract situation, because in most jurisdictions, the negative

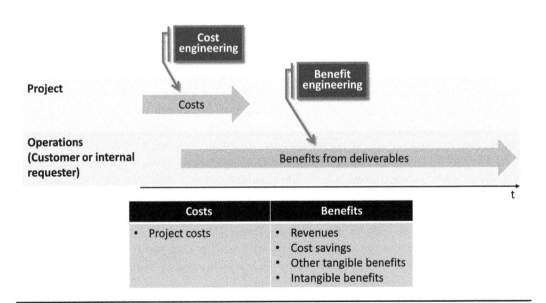

Figure 5-3 Focus of cost engineering vs. benefit engineering.

consequences are unforeseeable. While meeting customer requirements is an essential element of any project that is performed under contract, contractual requirements are the ones that the contractor first needs to consider. If they are met, the customer will probably not be delighted, because they are just expected; but if they are not met, the customer will be angry, and you know that he or she will be right. Meeting a contractual requirement does not make you look shiny, but not meeting it can send you into deep difficulty. In internal projects, the project is not performed against a contract but is an internal agreement. If the organization takes internal agreements seriously, the same applies, except that conflicts will not be remedied in a court but in the board room. In some companies, this option may be even more uncomfortable.

Another limitation for a project can be meeting legal requirements, such as deadlines and functional requirements for the implementation of software for tax purposes, or a waste water treatment system that is a mandatory prerequisite for production to be continued, or something similar. Making such a project less expensive by reducing scope or slowing down some activities could at some point cross the borderline of legality, and most organizations wish to act in compliance with the laws of the countries in which they are active. If you are performing a customer project, the law will probably not apply to your own organization but to its customer, and next to making profit for your employer company, taking care of the customer and giving them strong evidence that they selected the right vendor when they awarded the contract to your company is probably also part of your assignment as a project manager. In addition, customer projects are performed in a much more public environment, and whereas top managers in a failed internal project may prefer keeping it secret, a dissatisfied customer may well talk to the press or other media about the poor performance of your organization and destroy future opportunities for your organization to win bids for the rare project contracts that are truly attractive from a commercial perspective.

Figure 5-3 also shows the focus of benefit engineering: increasing the value for the internal requester or external customer from the use of the deliverables of the project, from its products, services, or other results. Often, but not always, it is a way to get a project out of a crisis, which is often overlooked. The customer may be prepared to talk again about price, deadlines, functionality, disruption of operations, and other difficult topics, if the contractor has something special to offer. The same is true for internal projects in which the project manager needs to talk with top managers about the budget, dates, and other key data of the project.

The most important prerequisite: You must know the key stakeholders who will make the decision, or at least influence it, and understand their wishes, visions, concerns, and what keeps them awake at night. You must also know the constraints that can impact their ability to comply with your ideas. You are risking a rejection, but compared to risking a crisis in your project, this is the smaller risk.

For many project managers and their teams, engineering benefits is a rather easy task. In their projects, they touch things that have not been touched for a long time. They see inefficiencies, inconsistencies, safety hazards, and lost opportunities on the customer or requester side, which often no one else sees—at least no one who is interested in fixing the problems. You should never assume that a company, a government agency, an association, or any other business organization is perfectly organized and managed. Most are loose collections of processes, some of which are working very well, while others are deeply flawed. Legacies may be a cause for such inefficiencies, old systems that work well enough to not put the organization under pressure to replace them with new ones. It may well be that no one has the audacity to touch them, because no one can say for sure what consequences such action will have. They have been around for a long time, and whereas people in the organization know how to use them, they do not know for certain what these systems actually do. Workarounds and improvised fixes are another common problem: They have been used in the past to provide temporary solutions to problems with the intention to correct them thoroughly later, but attention moved away from them and turned to more urgent issues. The solutions worked sufficiently well, and the company could afford the additional costs and efforts that they often caused. It may well be that the people who implemented the solutions are no longer in the organization, and the younger staff members think that the workaround is the way that things should be done*. An old proverb says that nothing is as long-lasting as a makeshift. When organizational leaders think of fixing these organizational or technical causes of material and immaterial waste later—something they actually may consider important—other urgent things occur, and the urgent is always the greatest enemy of the important.

Then, someone may benefit from inefficiencies. This person may be the only one who understands how to use the workaround and considers this monopoly of knowledge a job guarantee and a reliable source of power and influence inside the organization. Whereas they publicly pledge to stand by their work, they rather expect the work to stand by them. It is important to identify profiteers of corporate inefficiencies and improvised fixes to predict resistance that they may bring. Temporary solutions add a lot of complexity to systems, and this benefits those people who have learned to handle and manage this complexity. A StaFFA, as discussed above for the benefit engineering activity, developed jointly with a knowledgeable focus group may give interesting insight into the driving and restraining forces that the project manager and the team may face when they try benefit engineering.

A common cause for the long-term survival of inefficiencies in many organizations is the compulsion for them to operate without interruption. One of the pillars of the Toyota Production System (TPS) is called "Jidoka". Jidoka has contributed massively to the corporation's success over the last decades. The rule is quite simple: it says that if there is an error found in production, the line is stopped. This is a highly expensive policy, and the cause of the stoppage possibly may be a marginal mistake, but the effect is strong. The entire production line waits, and the attention of all involved focuses on the location of the error; all workers and managers of work stations are encouraged to consider whether they contributed something to the problem. The intention is not only to fix the error, but also to identify and remove the cause or causes and make the line more reliable this way†. Without Jidoka, the cause for the error remains in the line, and the error can re-occur. In Toyota's understanding, every error is an opportunity to remove an inefficiency or hazard from production, and this opportunity is used to increase efficiency—a fundamental asset of any production.

There are organizations that cannot stop their operations as "easily" as Toyota can. The social network, Facebook, had to stop its services in September 2010 for more than two hours. The explanation‡ given was technical but included the statement that they had to resolve a stable problem by turning off their site, something they considered "painful". They also

---

* This is similar to project management, in which many methods and tools that we know today were originally workarounds of problems with other methods and tools.

† A brief description of Jidoka and many other elements of the TPS can be found on the website of the car manufacturer (Toyota, 2015).

‡ (Johnson, 2010).

conceded that in their other systems, design patterns were used "that deal more gracefully with feedback loops and transient spikes". Facebook cannot simply turn off its services when a problem becomes visible and resolve the root cause. Interruptions of services are a major calamity for the company and also for the millions of websites worldwide that use their services. The same is true for most providers of Internet services, from flight booking to online learning to web shops and also for others conducting continuous operations such as suppliers of electricity. Their need to deal with errors without interrupting production often leads to a progressive cumulation of workarounds and unresolved root causes of problems.

I had an opportunity to perform seminars at an Internet flight booking provider and saw their ship diesel engines in the basement that protected them effectively from major power outages. However, a small, seemingly insignificant leap second inserted into Universal World Time (UTC) during the night of June 30 to July 1, 2012, brought its services to a halt, which had immense implications on flight booking internationally. They were not the only online company with such problems that night. Again, it is normal in uninterruptable systems that insufficiently handled causes of errors accumulate over time and develop the potential to wreak havoc in the organization and its environment. It is also a normal process that project managers and their teams stumble from time to time over such problems, and that they have the understanding and the resources to fix them.

Project managers may also stumble over opportunities for improvement. There may be unidentified and unexploited enablers in the organization, which may add tangible business value and make business managers shine in the perception of their key stakeholders. Project managers may be the only people to identify these opportunities, because during their projects, they become more and more insiders in the organization and gain the trust of their own management and that of the customer. They can propose value-adding changes to the project that are linked with decisions on new delivery dates, increased price and costs, and whatever additional burden the project will pose for the customer or requester, with the objective of adding business value for the project for all key stakeholders. Their work is intended to bring change to the organization anyway, which comes with some fundamental uncertainty, so the interest in listening to their recommendations is there.

Benefit engineering can combine the two approaches: fixing deficiencies and creating delight. It has technical as well as monetary and humanist aspects. Applied correctly, it is deeply ethical, because it offers the internal requester or the customer a benefit for the resolution of problems, often homemade ones, and these stakeholders have the right to say "No". Sometimes, it may even be useful to communicate the proposal in a fully transparent way: "Dear customer, we have a fundamental problem with the delivery date (or price or something else), but while looking for a solution, we found one that you may especially like". You have to know your stakeholders to predict if they will react positively when you use such an approach.

Benefit engineering may help the project manager shine as well. The project is not in trouble, but benefit engineering may increase its monetary or strategic value or create solutions that excite stakeholders. It may lay the foundation to turn a good project into a great one, a true "Wow Project"[*].

In customer projects, an interesting aspect of benefit engineering is that the proposal submitted to the customer is not developed in a competitive situation. In most cases, it would be

---

[*] A term probably coined by Tom Peters (Peters, 1999).

unlikely that another vendor would be able to make a competitive offer: The incumbent contractor has resources and infrastructure dedicated to the customer, possibly at the customer's location, and has much more detailed knowledge about the situation than any competitor. In some environments, this may lead to a problem for the customer-side managers, who must consider procurement procedures and limitations on business that may be awarded without competition, especially in public environments, where these limitations may be imposed by law. When the decision is made to attempt benefit engineering, all these factors need to be considered.

There are some more caveats. One is that the desire to rethink delivery dates, prices, or budgets and other key project data may collide with the inevitable constraints of the stakeholders' reality. A legal or contractual deadline cannot be simply ignored, more budget or resources may not be available to these stakeholders, and they may have to consider more factors that limit their freedom. There is no benefit in proposing changes with which the requester, paying customer, or other person disagrees. The project manager must know these limitations before such a proposal is made.

Another point for consideration is the generally two-faced nature of change control or change request management. A project in crisis can be salvaged by intelligent changes, but a project may also be driven into more difficulty if change control is not performed with sufficient diligence and care. Later in this chapter, I will describe what protective change request management should look like. One should also make sure that the proposal is submitted to the decision makers in a way that they find acceptable and that avoids giving them a feeling of being outsmarted.

Benefit engineering to salvage or polish a project is not a risk-free approach. It is recommended to consider the factors shown in Figure 5-4 when the proposal is developed. In a customer project, it has the same risks as any other business proposal to a buyer.

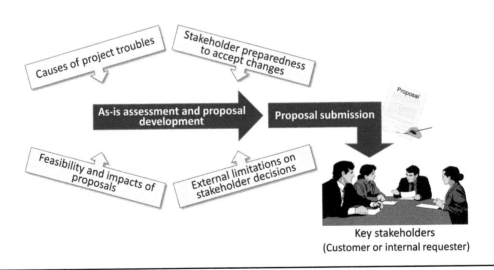

Figure 5-4   Considerations for the development of a benefit engineering proposal.

5.4 Pressure-Free Estimating

One of the most difficult tasks of a project manager is knowing what is going on in the project. One of the factors that make this task so difficult is pressure. Pressure is often applied to teach people lessons of compliance to imposed rules but ends in a *de facto* instruction on the most effective ways to circumvent these rules, to hide important but uncomfortable information from surfacing, and to obstruct the execution of directions from project managers with whom they disagree.

During a coaching and training assignment at a software company that I will call the Lizard Corporation, I saw how estimating can go wrong. Carl, a software engineer, had been assigned a task in an internal project and was asked by the project manager, Lisa, to estimate on how long it would take for him to finish it. After some time of reluctance and unwillingness, he finally came up with a number: two weeks, or ten work days. He had to wait for a colleague, who had to collect and engineer the requirements of the solution; and based on the results of this work, Carl could develop the basic architecture and the internal processes of the software. Lisa, the project manager, expected the results two weeks later to allow developers to start implementing Carl's design and develop the software. The developers were booked and ready to start working, but Carl's design was not. Lisa talked with Carl, and it was not a pleasant discussion with him in my presence, and when he had to concede that he was late, she asked him: "Didn't you promise me two weeks? I relied on your estimate and planned the two weeks for the project accordingly". His answer was: "No, I did not promise anything. The two weeks were an estimate based on the information that I had at that time. Today, I know more, much more".

Estimating is obviously a common source of conflicts. From the estimating party's point of view, estimating is considered a process that deals with uncertainty. From the point of view of the recipient of the estimates, the estimator is expected to understand the estimates as some kind of commitment, because the estimates are used to develop plans.

Estimates are done in a variety of areas in which a project has something that can be measured. They may refer to technical details, effort, duration, costs, times of disruption of operations, and many other things. Conflicts may occur with all of them. An estimate is an assumption with a number, and it shares uncertainty with all other assumptions. In other words: the estimate may be wrong. The uncertainty of estimates is often described in statistical terms such as Sigma (σ) and Variance (σ^2)[*], but project managers should bear in mind that the application of mathematics here is statistoid. The typical output number in a project is one, and this makes the application of statistical methods at least questionable for a major part of our work. We nevertheless often use statistoid methods with the simple excuse that we have nothing better, and often enough, they work surprisingly well.

Estimates are the basis of planning, but also of reserves that are put in place to protect the project from the uncertainties that are a natural element of any estimates. It is normal that durations, costs, and all the other estimated dimensions come out a degree under or over the estimate. The problem in the example with Lizard above was that the project manager, Lisa, asked for just one estimate and expected Carl to precisely meet the date. Carl had many other things to consider. The basis of functionality, performance, manageability, and options for

---

[*] A Gaussian term, not to be confused with the word variance used to describe the difference between plan data and reality.

further development of the software that his design brings, the risks that may be inherent in his design that may later jeopardize the software, the interfaces with other software in the organization that may cause entire systems to break down if they are not properly developed and implemented, and many more. Even if a project has very good engineers, developers, and other kinds of motivated, creative, and focused team members, project success depends on their motivation, creativity, and ability to focus, and it is rather unlikely that activities will end precisely on the predicted date. Sometimes, when work is easy, the process may be faster; other days, getting the work done takes longer than expected. Sometimes, private problems may interfere with work.

I visited a software company recently in Silicon Valley[*], which had pre-pasted disposable tooth brushes in the restrooms. The company found that software created by developers with achy teeth had lower quality. The company also reduces absenteeism with this inexpensive measure. For operations, it is much easier to prevent personal matters from impacting work; for instance, a love-sick bus driver is normally quite capable of driving the bus. In operations, a lot is done by routine and also by habits that develop over time. In projects, it is far more likely that personal issues will affect the project and cause delays and distress. I would never expect a love-sick software developer to finish work on time.

Other factors influence how people will meet their estimates: The dual or even multiple reporting relationships in modern matrix organizations, the dependencies on the timeliness and quality of predecessor work that first must be done, the uncertainty of other work that the person has to finish and that blocks the person's availability, changes in the internal and external business environment of the performing organization, and many more.

Another aspect that makes estimating difficult is the membership of the estimator in one of two factions that we commonly see: the faction that believes in the *"panic"* monster, and the faction that is trustful in the *"easy"* monster. The panic monster makes people estimate overly pessimistically. The estimates are political estimates that are easy for the person to bring to fruition. Their time estimates have a lot of padding, and their cost estimates include major monetary reserves. Pessimistic estimating may also be a sign of deep politeness: "I do not want to disappoint my project manager".

On the other side of the strait sits the easy monster. There may be more than the three groups that I know on the easy side, but I found mostly sales people, politicians, and young people fresh from university exaggerate on the optimistic side. These are often people who are afraid that the task will be given to someone else with more optimism, or may be dropped because it takes too long, is too expensive, or poses too much of another kind of burden on the organization. As project managers, we understand that we must often navigate between these two monsters, and this is the case here. We do not want political estimates, what we need are just estimates. Somewhere between the two monsters is realism, and this is the route to take. Project managers have to take politics out of the estimation process because politics drive estimators either to the easy or the panic monster.

There are several factors that influence estimating and make it political, and there is a booster that increases their effect. One factor is pressure, which has many negative effects. It tends to lead people into game-theoretical dilemma situations. Experience is an important

---

[*] It would of course be interesting to know the name of the company, but I am under nondisclosure agreement (NDA). And to respond to a hygiene concern that I sometimes hear when I talk about this observation, the tooth brushes are individually wrapped.

element of project management, because it is a powerful teacher. However, people under pressure learn the wrong lessons from this teacher. Instead of the lessons of how to build strong teams, how to put the mission first, and how to cultivate an environment of trustworthiness and mutual empathy, people learn under pressure how to effectively obscure failures, finger-point blame to others, and sabotage production to get some additional work breaks—while their employers have to stop production and are desperately trying to get it up and running again. Pressure also leads to fragmentation of organizations. While it is important for managers to know what is going on in an organization, without a well-working secret service inside it, the level of knowledge for high-pressure managers is limited. Pressure can be successful for a limited time but cannot create sustainable systems. This is true for politics, where we see tyrannies successful for some time and then suddenly collapse, and we also see it in companies and in projects.

To illustrate what Lisa's approach did to the project estimates, picture a car and a narrow street in a little antique mountain village in Tuscany that a family needs to drive along to get to their summer vacation domicile. Most cars have a width including mirrors of just under 2 m (78.7 in). Some sports utility vehicles (SUVs), pickups, luxury cars, and sports cars exceed these dimensions by 10%. Imagine the distress that the driver feels in a 2-meter-wide car when the width of the alley is just 2.2 meters. The driver does not want to scratch the expensive car, so he or she must drive the car slowly through the narrow street, and every bend of the street and every uphill section increases the distress that the person feels. When the family takes the road for the first time, they may wonder if they are on the way to their "doomicile". People who live in Tuscany have a simple solution for the problem, they use small cars or mopeds that allow them to keep a comfortable distance from the walls to the sides of the street, but many visitors buy their cars based on other criteria, especially those who can afford Tuscany domiciles.

Carl, in the example, felt the narrow road to his doomicile as much as the driver with the wide car in the narrow road. When Lisa asks him if he will finish on time, as polite as he is, he will say yes, even in moments when it has become obvious to him that he cannot. Bob Wysocki called the effect "hope creep"[*]. This is when team members or contractors communicate that they are on schedule when they are actually late, but they want to avoid the pressure and hope that they can make up the delay with overtime and weekend work. Chapter 2 discussed how project managers learn. Hope creep is the attempt by certain stakeholders to prevent project managers from learning what is going on in the project; they are bending the learning curve in a direction that project managers learn late of delays and other problems, that they could manage easily if they got the information early, but meanwhile, they have run out of options and those remaining are more expensive. The game theoretical dilemma called chicken race, described in Chapter 2, is also a type of hope creep. Others must also be late in this project or program and it is hoped that they will report their delay first. Hope creep is the direct consequence of high-pressure environments, and in its wake, delayed projects with overrun budgets (see Figure 5-5 on following page).

The narrow Tuscany village road describes solutions when no alternative route is available: use a smaller car or widen the road. If the delivery date is near and inflexible, ask what needs to be done by that time and remove all unnecessary scope, at least for the moment. If the scope is inflexible, plan with schedule reserves. For both approaches, you need an understanding of the uncertainties that you need to expect, such as that you want the road to be wide enough

---

[*] (Wysocki, 2014a, p. 34).

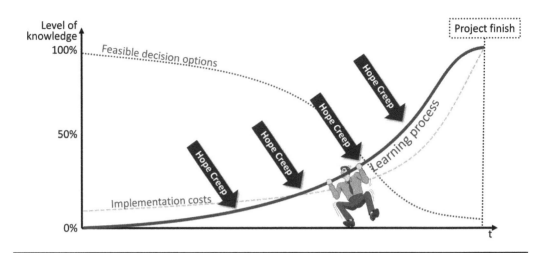

Figure 5-5   Hope creep bends the learning curve in the wrong direction and makes the necessity of decisions and the available options visible too late, when the number of options for decision making has diminished and their costs increased.

compared to the width of the car to make driving less stressful. For our project, we can achieve the same goal by defining objectives as corridors, not as single points.

The method of defining such a corridor based on estimates is called "three-point estimation". It is not new; it was part of the Program Evaluation and Review (PERT) network diagramming method developed in the 1950s and 60s for the U.S. Navy. The other parts of PERT have mostly vanished, especially their network diagramming using arrows to depict activities, but three-point estimating has not only survived, it has experienced some type of a renaissance. The basic approach is simple but powerful, exactly what project managers need. Instead of asking for an estimate, the project manager asks for three estimates and uses simple mathematics. Adding some psychology to the method makes it even stronger. The project manager asks for the three following estimates:

- P: the pessimistic estimate. If you are confronted with a lot of problems that can make the work difficult for you, what would be your highly pessimistic estimate?

- O: the optimistic estimate. If things go extremely easy and no problems occur whatsoever, what would be your most optimistic estimate?

- ML: the most likely estimate. Based on your experience with similar tasks, which estimate would you consider the most likely one?

In the following discussion, I will focus on the time aspect, but the technique is also applicable to cost, technical issues, work efforts, times of disruptions of operations, and more, as mentioned before.

The number used for further planning is calculated with a simple formula:

$$\text{Plan estimate} = (O + 4ML + P) / 6$$

To give an example:

- P = 13 days
- O = 4 days
- ML = 6 days
- PERT value: (4 + 24 + 13) / 6 ≈ 7 days

The results are also shown in Figure 5-6.

For the example, I rounded the value up. I would not plan to finish work on the seventh day after six hours and 40 minutes. This would communicate an accuracy that no estimate can promise.

A question that is often asked when I present the method in classes is, "Why do I use the weight of four that is assigned to the ML value?" There are various explanations why it is just four, none of them is fully satisfying. I recommend simply accepting it as a standard developed by the U.S. Navy half a century ago.

In reality, it is often faster to obtain the three values than just the single one as Lisa did. With three-point estimation applied with some empathy and intelligence, pressure on the estimator is reduced and so is the person's fear. In addition, the resistance against estimating decreases and the numbers are communicated much more quickly.

The approach one should take should be different for optimistic and pessimistic estimators. Let us first look at how the method would be used with pessimists. Their fear is that their estimates are too optimistic and are then turned into commitments, so take the pressure from them by starting with the "P" value: "If things go very, very badly, what duration should I expect?" As this is the value that they have mostly in mind, they are likely to communicate

3-Point Estimation
(From PERT – Program Evaluation and Review Technique)

1 * O	4	4
4 * ML	6	24
1 * P	13	13
Sum		41
PERT weighted average	/ 6	≈ 7

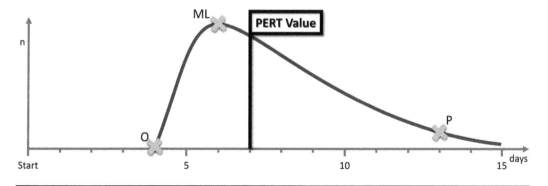

Figure 5-6 The assumed probability distribution of durations for the estimated activity in the example.

it rather quickly. It is important that this value is documented to take the pressure from the estimator when more optimistic values are discussed. Then, the question turns to the "O" value—which is now much less of a problem for the person—the possibility that the activity may take longer than what already has been communicated and documented. It may be helpful to remind the person of the earlier days, when we were all much more enthusiastic and loved to make estimates that were almost impossible to meet. The third value is then the "ML" value, which should be located between O and P, mostly nearer to the O value, by the way.

The approach is of course again statistoid. The activity is done once and not repeated in major numbers to verify the correctness of the assumed distribution model and the individual estimates. However, it works well in practice.

Optimists feel a different kind of pressure: Someone else may get the job who promises faster delivery, or the activity, or even the entire project, may be considered to be too time-consuming and expensive and is terminated. To take pressure from them, the first question should be for the optimistic value, the one they like most, then the pessimistic, and the most likely value. Once the favored estimate has been made, fear is much lower to add realism to the estimates and prepare the pessimistic and the most likely estimates.

Sometimes, estimators feel uncomfortable about the ML estimate. For such cases, you can use a two-point estimation approach by calculating the plan value with the following formula:

$$\text{Plan estimate} = (P - O) / 3 + O$$

This computes for the example:

P = 13 days
O = 4 days
Plan value: (13 − 4) / 3 + 4 = 7 days

This two-point estimation is less accurate but is a second-best solution and sufficiently accurate in most cases. The plan value lies between the values of O and P, nearer to O than to P, and this is often where the PERT value arrives, especially when some rounding is applied.

Three-point and two-point estimates have a number of benefits over the single estimation.

- Two-point and three-point estimates create a result corridor. The activity is expected to be finished inside the corridor, and the project manager should prepare the project for any result inside the corridor. This is a better planning basis than a questionable commitment on a certain date.

- The areas under the distribution curve that are right and left from the plan value are roughly equal. This means that the likeliness to stay under the plan value or exceed it is roughly 50/50. To be more accurate: If the activity would be done repetitively, Carl would often be faster than planned, but when he is slower, he may be much slower. This even distribution allows for robust estimates in which over- and under-estimations balance out. The project manager then needs a set of suitable planning methods to use the first to cover the second. This is easy with money, as a dollar saved in one activity can be used to support another one that has been found more expensive than expected. Money is universal. It is more complicated with work effort. The team member that unexpectedly finished work early in one activity may not be qualified for the work in another one. It is especially difficult with time, because the flow of work may not allow an activity to be started before its predecessor activities have delivered their outputs, which this activity needs as input.

- The project manager receives a measurement of the uncertainty that comes with the activity following the formula P – O, in this example nine days. This uncertainty is one of the two important pieces of input to calculate reserves. The other input is criticality. When a project manager has an appointment with a trainer, such as taking me to dinner, the person may be fairly late. The same person is likely to be early for a meeting with a recruiter who has a job opening. If the journey time to the meeting is uncertain, for example, one that is due to uncertain traffic conditions, the person will leave earlier. If the time is less uncertain, the person may, for example, be able to walk, and the person will leave much later for the meeting. The two influencing factors both play a role, and for the uncertainty, P – O is a great metric.

- The project manager can develop an early warning system for problems with the activity by asking repeatedly for new estimates for the remaining work or duration. The difference P – O, the metric for the uncertainty, should decrease over time, as the team member should become more secure with the progress. I recommend using candle diagrams to visualize how the uncertainty decreases. Figure 5-7 shows two activities. At first glance, Activity #1 seems more difficult; the PERT value is going up, which means that the task will come in almost four weeks later than originally expected, but still inside the corridor.

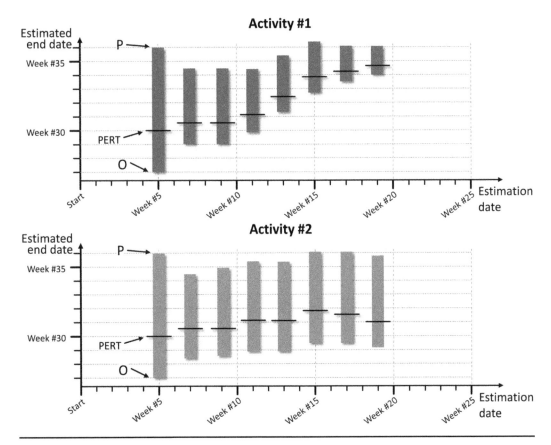

Figure 5-7   A candle diagram visualizes that the team member performing Activity #2 does not gain certainty, while the person proceeds with the activity. This is a warning sign of difficulties that have not been communicated so far. The project manager should ask what the problem is.

Activity #2 has lost only one week during its advancement. The candle diagram, however, also shows the team member performing Activity #2 has not gained any certainty in the estimates, which is a signal the team member may have some kind of problem. If the problem has not been communicated, this is a good moment for the project manager to ask so he or she can know about the problem earlier when more feasible options for decision making are open and when the costs for implementing them are still lower.

There is another question from the initial exercise that needs to be answered. If Carl gave a one-point estimate of 10 days, and the estimation points for the activity were O = 4, ML = 6, P = 13, and PERT = 7, the likelihood for Carl to meet the ten-day delivery date should have been very high. Why did he miss it? There are several explanations, including the probability distribution along the PERT values, which goes to 13 days, and the 13 days are not yet a worst case, just a pessimistic estimate. Here is another explanation. Carl knew, from experience and intuition, that it would take him about seven days on average to do the activity. He gave himself ten days so he felt the activity is not urgent. It may be important, but it is not urgent. Then, he had another task to perform, which was urgent but was not expected. The urgent is always the greatest enemy of the important. The urgent task would have normally taken him two days to finish, so it would probably not collide with the activity in the project, as long as it is not one that is too difficult and tedious. So he decided to do the short, urgent task first. Maybe it was not his choice to make this decision but his line manager's decision. Then, a problem occurred during the task, it took him more than the three days to finish it, and he started the activity for Lisa's project late. Then we assumed some problems occurred in this task as well.

When Carl padded the estimate for the activity, he took the pressure away from it but also the sense of urgency. The tailor rule[*] says: "If you cannot be on time, be early". The art of three-point estimating is to take the pressure from the schedule without making things look less urgent. In the example, we would expect Carl to deliver in seven days, but we would also be prepared in the not unlikely case that it takes him longer. As the method leads to shorter activity durations, the time saved there can be used to protect the constraints of the project such as deadlines, funding limitations, and others by reserves. If we give up the sense of urgency, our project will always be late.

Is being early a waste of time? Not necessarily. Tell your team members to use the free time to prepare themselves, read documentation, talk with colleagues or the client, and do all other things that help them start the next activity when the time comes, without delay and as planned.

5.5 Protective Change Request Management Process

Eighteen kilometers (18 km, 11 miles) south of the city of Berlin is an airport called "Flughafen Berlin-Brandenburg Willy Brandt"—or to be more accurate, the construction site of the airport. Airport construction projects are notorious for massive delays and cost overruns, such as Denver Airport in the USA or London Heathrow Terminal 5 in the United Kingdom, but the Berlin Airport outrivals them all in this discipline. It was planned to be open in 2010, and while I am writing these lines in late 2015, there is no opening date fixed, but it is certain it will

---

[*] I call it the "tailor rule", because I learned it from Dave, my tailor. I am aware that he did not invent it, but he sticks to it.

not be opened in 2016. A vivid description of the history of the airport has been described by one of the architects of the airport building, Meinhard von Gerkan*. If your German is not sufficient to read the book, but you have a strong sense of *schadenfreude*—joy in the misfortunes of others—you may find an article in Bloomberg entertaining and informative[†]. There is a large amount of analyses available, and they have two failures in common: a dysfunctional fire protection system as a technical one, and ignorance of basic rules of change request management by the responsible people representing politics and management as well as personal ones. In project management, too often we meet people who I call Frog-king managers (covered in Chapter 3)[‡]. Instead of listening to their teams, contractors, and the experts, they are guided by the first sentence in the Brothers Grimm fairytale: "In olden times when wishing did help one . . ." For the dynamics of success and failure, these people can have a deeply disruptive effect.

I mentioned repeatedly the two-faced character of change requests. Wisely managed, one can use them to steer a project out of trouble, along with the benefit engineering tactic discussed earlier. Change requests can also drive a project into crisis. There are often some big egos in people working on a project, who see the project as a way to leave some kind of heritage to the world and build themselves a monument, and their zeal regarding what this monument must look like exceeds their understanding of what is manageable for a project by an order of magnitude. There are also negligence and lack of discipline by project managers and team members who do not understand that a poorly managed change request can turn into a time bomb.

What process can best protect the project from harmful change requests? First, we should have a clear understanding of what stakeholders in a project should consider a change request and what is not one. Chapter 4 contains a table with the distinction between refinement, making raw definitions more precise, and change. Both are part of progressive elaboration and therefore considered normal in a rolling wave approach to managing a project. This chapter refers to the second.

Project management deals with processes from different knowledge areas, as Figure 5-8 shows.

---

* (von Gerkan, 2013).
† (Hammer, 2015).
‡ As mentioned in Chapter 3.

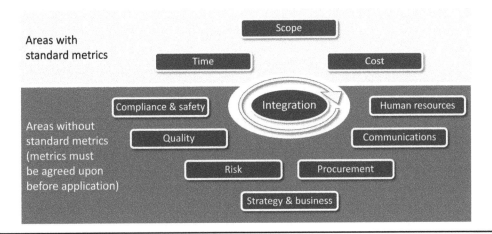

Figure 5-8 Knowledge areas structured by types of metrics in project management that should be considered in most projects.

Integration management is the knowledge area in which change request management should be concentrated. Then there are three knowledge areas with standard metrics:

- Project Scope Management, which includes work effort measured in units such as person-days, machine-hours, etc.
- Project Time Management, measured in time units such as days, hours, etc.
- Project Cost Management, measured in monetary units.

Sometimes, project managers deviate from these units—for example, when using the Earned Value Technique (a method to compare actuals with baseline data) the Schedule Variance is measured in dollars or another monetary unit. As long as people adhere with the standard units, no explanation is needed. Sometimes, the three areas (scope, time, and cost) are referred to as the "triple constraint" or the "magic triangle", and while there is a lot of discussion on what makes these elements of project management so special, it is obviously the existence of standard metrics.

In quality management on projects, one can also measure things, but there is a prerequisite. First, metrics—rules for measurement, including the measurement units—must be defined, which can then be used to quantify the quality dimension. The same is true for human resource management and other aspects of project management. As soon as metrics have been defined and agreed upon, one can start measuring.

When change requests need to be managed, all these areas must be considered, and in a situational approach, there may be more that matter for a specific project. Effective and protective change request management does not only relate to scope changes, but also to any change in the project, which potentially has an impact on the plans and their implementation.

Chapter 3 presented this approach to managing change requests. Remember, the appropriate response to any kind of change—including reduction of project costs, early release of team members, altering delivery dates, contractor changes, or whatever it may be that someone wants changed—is to open the drawer of the desk, take out the change request form, and ask the person to fill it in. Then the request will be submitted to an agreed-upon process. If discussions begin on your bureaucratic handling of the request, point to Berlin-Brandenburg Airport and obtain the support from your stakeholders so you can avoid such a disaster in your project. The change request process should not be blind bureaucracy but a protective mechanism for your project.

Who can request a change from the project? A change request can be directly submitted not only by the customer (if the project is under contract) or by an internal requester, but also by any key stakeholder. Indirectly, any stakeholder can cause the team to make changes to their plans and to the implementation, when this helps reduce resistance and increase support and engagement by stakeholders.

Next, decision levels should be agreed upon, as shown in Table 5-1, which is an example of an escalation matrix that may be useful for an internal project.

The example denotes another important requirement for the process: a defined entry point for the submission of change requests to collect them and separate those that have no chance of success from those that should be analyzed and considered as potentially beneficial. The change request management process should then be developed around the elements, as shown in Figure 5-9.

Table 5-1 Example of an Escalation Matrix for Change Requests in a Project

Entry Point for Submission of Change Requests: Project Sponsor		
Change Decision and Its Impacts	**Approval Level**	**To Be Consulted**
Small adaptations of product features	Team	Project manager
Changes inside buget, deadlines, critical requirements	Project manager	Experts, project sponsor, Change Control Board (CCB)
Changes that effect budget, deadlines, critical requirements	Project sponsor or CCB	Project manager, Portfolio Review Board
Continuation or termination of the project	Portfolio Review Board	Project manager, experts, project sponsor, CCB

(Escalation ↓ ↓ ↓ ↓ ↓ ↓)

Elements of the Protective Change Request Management process are:

- **Valid plans.** If the change request gets rejected, they remain valid. They should be consistent and compliant with the various project requirements and its deliverables. This includes staff members who are not booked during holiday absences or with assignments of over 100%. It also includes that no delivery is promised in six months, when work is necessary to achieve this time when it will take the team 10 months. And so on. A set of plans with inconsistencies and non-compliances means in essence to plan for failure. If everything will happen as planned, the project will flop. Whereas projects generally do

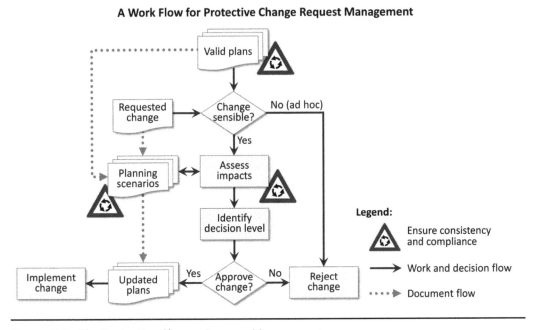

A Work Flow for Protective Change Request Management

Figure 5-9 The Protective Change Request Management process.

not perform precisely as planned, this part of the plan normally works. The entire process relies on the planning consistency and compliance with the plans. A consistent and compliant plan is easy to identify. If the project would be finished as planned, the project would succeed.

- **Requested change.** Submitted, ideally through the entry point.
- **Change sensible?** A quick check if the request should be rejected immediately because it is not plausible or unrealistic, or because it is too late. This step is important to avoid processing change requests in which assessing their impact is costly since they have no chance for success.
- **Planning scenarios.** There may be one or more scenarios; software can be helpful to develop a second or third scenario from a first one. These are project plans that are derived from the combination of the original plans and the change request and integrate the change request over all areas and consider constraints from the technical side in the same level of detail as legal or contractual constraints. On rare occasions, they may be entirely new plans. Planning scenarios are developed in sufficient detail and maturity so that they could be used as new plans, but they have not been decided upon and approved.
- **Assess impacts.** These are now simply the deltas between the scenarios and the still valid plans. This process step again requires an integrative view of the project.
- **Identify the decision-making level.** Based on the deltas and the escalation matrix, the appropriate decision-making level is determined, and the decision options are presented for approval or rejection.
- **Decision.** If the decision is made to accept a planning scenario as the new plan and with it the change request, the new plan is communicated to the team and implemented.

Are there exceptions or situations in which bypassing the process would be accepted? Yes. Field changes, also known as ad hoc changes, often needed in emergency situations, may dictate circumventing the process. They become necessary when human life or health is in immediate danger, the environment may be damaged, or when other threats have emerged. They are necessary when the risk of not acting or acting too late is larger than the risk of making the wrong decision. Constructive changes are another situation that may require circumventing the process. They are "field changes on steroids"—when the time is not even sufficient to formally agree upon the change, it is just done, and the actual change is later construed by the parties or in the worst case by a court into what was contractually agreed upon and what has actually been executed. Field changes and constructive changes are often done in panic situations and have the potential to damage a project immensely because there is no time to develop the impact analysis and make an informed decision. It takes a lot of experience and discipline to implement them to the benefit of the project.

5.6 Registers

In earlier decades, project managers kept a set of "logs". Later, the logs were referred to as "lists", and then they became "registers". In essence, they are all the same, and while keeping registers seems inexpensive and trivial, they are inexpensive but *not* trivial. One does not even need expensive software for them, just anything that enables building tables. Paper and pencil are also fine.

5.6.1 The Assumptions Register

The worst industrial disaster that has struck Japan since the Second World War was the destruction of the nuclear power plant at Fukushima Daiichi on March 11, 2011, by a magnitude 9 earthquake, followed by a 14-meter (15.3-yard) tsunami that flooded the entire area. The accident was rated at level 7 on the International Nuclear Event Scale—this is the highest rating on the scale. It consisted of three meltdowns, and in terms of nuclear disasters in peace time, the technical havoc could not have become worse.

Fukushima Daiichi is located in only 225 km (140 miles) northeast of the city of Tokyo and the surrounding Greater Tokyo Area, where over 37.5 million people live—more than the population of Canada. The area is one of the most populous and also most densely populated metropolitan regions of the world. One would expect that the engineers, who built the power plant and commissioned it in 1971 in the proximity of such an area, would have dedicated an intensified degree of diligence to the safety of the plant, given its closeness to such a vast number of people and also given the frequent maritime northeast winds that could carry nuclear fallout from the plant to Tokyo in just hours. The events in 2011 have shown that this is not what they did.

Reports on the disaster point to a root cause: assumptions turning into facts (see Chapter 2). Simulations that were made to test the robustness of the plant in the case of seismic incidents assumed a much smaller earthquake and, therefore, a much smaller tsunami as a consequence. The subsequent results of the research were acceptable to the engineers. It saved them costs, it simplified the project, and it accelerated the entire progress. The *Wall Street Journal* reported in 2011 that the location of the Fukushima rector was originally 25 meters higher, well above the height of the 14-meter tsunami, but in order to save construction cost and get easy access to cooling water, the ground was broken and the plant was lowered to a level of ten meters, which the tsunami easily overflowed[*]. Discussions on the questionable basis of the assumptions were not allowed, and diverging opinions were suppressed[†]. With over 35 million people living in the vicinity, such behavior is not a sign of responsible decision making, as history showed in 2011. People in the Greater Tokyo Area were lucky that the winds were blowing most of the radioactive fallout over the sea, so that the landside contamination remained low except in the direct neighborhood of the plant.

Independent of the industry, many project managers have experience with factoids. When I address the topic in class, the majority of my students have similar stories to tell. Assumptions turned into factoids are often communicated in statements beginning, "We all know . . ." or "There is consensus among experts . . ." No one verifies them, because no one wants to hear the uncomfortable truth. Another aspect is that it is much easier to formulate a hypothesis than to verify it, and most project managers are paid to deliver, not to do research. So, for the

[*] (Dawson & Hayashi, 2011).
[†] (Acton & Hibbs, 2012).

sake of simplicity and to please key stakeholders, people often assume that assumptions are not assumptions. Factoids can directly damage the project or, as happened in Fukushima Daiichi, result in time bombs that go off years later.

One word on time bombs: It is interesting how people can get used to active time bombs. In action movies, they are often used to put a hero under pressure to make a decision if he or she will cut the blue cable or the red cable to disable the bomb. In these movies, the ticking clock always shows the time when the bomb will go off, there are only seconds left, and this adds significantly to the thrill for the audience. The audience has paid for the thrill, so it is a perfectly acceptable approach in entertainment. Reality is not meant to entertain, it is just there, and it is different. There is no ticking clock and no display, so people are calm when they should be alert. For them, protecting oneself, their loved ones, the community, and the environment from a known time bomb may be important, but there are more urgent things to do today, and the urgent is at all times the enemy of the important.

There should be a major earthquake in Southern California every 110 to 140 years. The last one was in 1857, almost 160 years ago. Some people take this as calming observation—the risk is smaller than it was in the past. They should not be calm. Tectonics in geology follow a rule that if an earthquake has a historic repetition frequency and comes late following this frequency, it will is likely to be more vicious. The area at stake has over 22 million inhabitants and includes the cities of Los Angeles and San Diego.

Another example: Mount Vesuvius in Italy is an active volcano that had a small eruption every 13.5 years between 1660 and 1944. In the 70 years since the last eruption, it has remained silent and has probably built up internal pressure that will make the next eruption more vicious. It has a notorious history of destroying villages and cities with ash rains and pyroclastic flows, highly mobile clouds of hot gas, and molten ash and rock that come down the flanks of a volcano with speeds of up to 700 km/h (400 mph). The most famous of these destroyed the cities of Pompeii and Herculaneum in 79 AD. Today, the modern province of Naples, with three million inhabitants, is in the direct vicinity of the volcano; it is Italy's most densely populated area. The distance between the peak of the Vesuvius and Naples main station is only 13.5 km (8.4 miles), and one can easily see the volcano from inside Naples. Scientists found a 4-meter (4.4-yard) thick layer of volcanic matter under the city, explained by an eruption 3,800 years ago, that was far more devastating than any other known in this area and that could repeat at any time. Vesuvius is a permanent threat for everyone who lives there, and it is seemingly hard to ignore, but this is what people do. They feel calmed by the statements of scientists that the next eruption will not be as disastrous as the historic ones, and there will be enough time to evacuate Naples and the surrounding cities if the volcano becomes active. An evacuation plan has been developed, which adds to the sedation. This is all based on many assumptions: one that there will be no panic that would cause people to clog roads and trains; another that there will be enough time to execute a decision to effectively evacuate the region—a difficult decision given the risk of a false alarm; and a third that the eruption will not be a surprise but will be preceded by early warning signals. Another assumption is that the wind will not blow clouds of ashes in the direction of the city of Naples. Comforted by scientists, the people from Naples have become used to the time bomb nearby and live their normal lives. When one talks with them, they are more concerned about the urgent deficiencies of their public services and traffic conditions, caused by a widely dysfunctional administration (as they say) and the throttlehold of the Camorra, the local organized crime organization that debilitates the society there. Many of them see Vesuvius as a source of income; it is one of the greatest tourist attractions of the city. However, one day, the time bomb will inevitably go off, and the past destructions of Pompeii and Herculaneum give a strong indication of its devastating power. Recently, a small number of scientists recommended completely evacuating a region of 20 km around Vesuvius, which would include the entire city of Naples[*]. This recommendation is of course also based on assumptions, but these have been developed with more accurate data. In Naples, people do not discuss the recommendation and would probably never follow it.

Back to project management, where most disasters are fortunately much smaller and less devastating. Assumptions are a necessary element of project management. Many decisions must be made early, before sufficient knowledge is available, and corporate managers, project managers, and team members will have to make assumptions and then base decisions on them. Assumptions are made during the internal project selection process or during the business development process in customer projects. They are later made when the project is chartered and planned and accompany executing and monitoring/controlling. Making assumptions is sometimes even necessary during the formal close-out of a project.

Assumptions may be wrong.

How can project managers inhibit the development of factoids from these assumptions that ignore the natural uncertainty of assumptions? How can they avoid having assumptions calm down stakeholders and make them ignore time bombs and other threats that should be studied early and considered in decision making? A very simple solution: make a list of assumptions, or in modern terms, prepare an assumptions register. Whenever an assumption is made by the project manager or by another stakeholder, it is added to the register with some basic information: when it was made, by whom, and—the most important information—what decisions were based on it. Eventually, the assumption is replaced by evidence, which confirms the assumption or replaces it with better knowledge as part of a normal learning process, as described in Chapter 2. The project manager then marks it as closed but keeps it on the register for later evidence. There are three times that such a register will prove to be helpful for a project manager:

- Someone turns the assumption into a factoid, such as: "Scientific simulations have shown that a tsunami cannot be higher than five meters at this coast". All the project manager needs to do is point to the register and ask if evidence has been found that proves the assumption. If this has not happened, the assumption remains what it should be: an uncertainty with potential effects on the project and its deliverables.

- An important process of project risk management is risk identification. The assumptions register provides valuable input that can be used during a risk-finding workshop. The attendees walk through the register, entry by entry, and ask about each of them: What if the assumption is wrong? Further questions may be: Why could it be wrong? To what degree can it be wrong? When can we know for certain if it is right or wrong?

- Decisions are often criticized by stakeholders with knowledge from hindsight. "How could the project manager make such a poor decision?" The assumptions register may then help remind people of the inadequate information available at the time the decision needed to be made.

---

[*] (Barnes, 2011).

Another issue in project management is rework—the time that project teams spend working again on deliverables that have previously been considered finished. When assumptions prove wrong, a common necessity is rework, which causes additional availability of resources to become necessary, together with adjustments to the schedule, causing the budget, operations, and other projects to be be disrupted, and so on. An assumptions register helps stakeholders stay prepared.

The assumptions register is a great tool for SitPM, as it helps the project manager and the team not only develop a better understanding of the relations between uncertainties and decision making in specific situations but also remain concerned about its risks. It is also a continuous documentation of the learning process and how the team develops certainty out of uncertainty over time.

5.6.2 The Constraints Register

Constraints are hard limitations. Many constraints are found or imposed at the onset of the project, while others occur as the project is planned and executed. Deadlines are constraints and so are budget and funding limitations. There may be technical constraints or others relating to resource availability. They all have in common the fact that the project manager cannot decide upon them, while they limit the project manager's freedom to make decisions.

Constraints have three concerns for project managers:

1. The project must meet them. This is simple of course, but keeping track of how a project performs against a multitude of constraints can be a difficult task. An example: A project has four deliverables that must be finished on time and that have to be created in four sequential activities. Deadlines have been imposed, and durations have been estimated as follows:

   Project start: November 2
   Activity A. Deadline: November 30, duration 20 days*
   Activity B. Deadline: December 24, duration 17 days
   Activity C. Deadline: January 7, duration 15 days
   Activity D. Deadline: February 5 (= overall deadline), duration 14 days

   Such a list is quite confusing, and this is still a simple one. Each of the four duration deadlines must be met. At first, a brief overview may look promising. After more detailed analysis, however—for instance using project management software—you see that one deadline cannot be met, the third one. Observing one deadline may be easy, but observing a multitude can be difficult. Project management software in the hands of an expert can be helpful for such tasks. The constraints should be listed in the constraints register so you do not forget one of them or lose track of them during the course of the project. Not realizing that the project is on a path to miss one or more of the constraints can be detrimental to the project. You should use scheduling methods and possibly software to identify those constraints that you are likely to miss. Figure 5-10 shows an example of

---

* To keep the example simple, days means working days, a week has five days, and we ignore holidays.

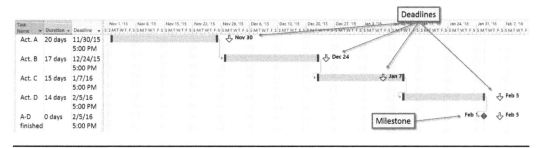

Figure 5-10 The project has no difficulties meeting the first and second delivery dates, but the third is a problem.

how network diagramming done with project management software can help identify the truly difficult deadlines.

2. Figure 5-10 shows another effect of constraints: They are part of the identification of reserves in time, cost, and other dimensions. The milestone "A-D finished" should be completed on the evening of February 1. The deadline, a time constraint, is set for February 5. This leaves a schedule reserve of four days. If something unexpected happens, these four days may be used to respond to the problem. What are reserves? Reserves in project management are generally defined as the difference between objectives and constraints, or the differences between what we want and the limitations of what is or is not allowed. Reserves can be "slack" or "float"*, if they are the incidental results of the logics in the plans, or "buffers", if they are deliberately placed by planners to protect the project from missing the constraint. Excessive buffers are sometimes referred to as "padding"; as most reserves come with costs, padding can be quite expensive. Sometimes the project lacks any reserves, which means that the project can be derailed by minor unplanned events. Table 5-2 (on following page) is an overview of the generation of reserves from objectives and constraints. A project manager without reserves is a feeble bystander of important events.

3. A constraints register is a valuable tool for risk identification along with the assumptions register. When the team becomes aware of a constraint, it is often unclear as to its impact on the project. One constraint may not have an effect, while another one may make it impossible for the team to meet all requirements and can derail the entire project. During the risk-finding workshop, the constraints register can be used as a second source for risks. One discusses them one after the other and asks, "What risks can each engender for the project?"

5.6.3 Requirements Register

There are various approaches to listing requirements, from using simple checklists to a "Requirements Traceability Matrix" to a more complex "Requirements Breakdown Structure"†.

---

* Mostly used for schedule reserves.
† (Wysocki, 2014a, p. 158).

Table 5-2 Reserves Are the Difference Between Objectives and Constraints

Area	Objective Statement	Constraint Identified	Reserve
Time	Start of Production (SoP) has been targeted for January 1, 2016.	SoP deadline has been imposed for April 1, 2016.	Three-month schedule reserve.
Budget	A cost estimate at project start has been approved: $10,000,000.	There is a funding limitation of $12,500,000.	$2,500,000 monetary reserve.
Scope	15 functions are planned as "wanted".	12 functions are specified as "critical".	Three nice-to-haves (reserve for de-scoping).
Quality	Control limits: Dimension x ±1 mm.	Specification limits: Dimension x ±1.25 mm.	Quality control reserve: 0.25 mm[1].
Resources	10 team members have been planned.	12 team members are available if needed.	2 bench resources.
Resource availability	Bill is expected to work four weeks on the task.	We have booked Bill for five weeks.	One week resource reserve.
Operational disruption	We plan to stop production for four weeks.	Management has agreed to a production stop of a maximum five weeks.	One reserve week for the project.
Procurement	The contract has been awarded to a contractor.	A second contractor is on standby for a fee.	One backup contractor.
	The delivery has been ordered for April 1.	The delivery will be needed for the project on April 8.	Eight days feeding buffer.
Risk	The work package is planned for $10,000,000.	Insurance has been bought to hedge the project from losses from certain risks.	The benefit of the insurance[2].
Safety	The power plant must be protected from a 3.7 m tsunami.	The power plant has been built on a location 10 m above sea level.	6.3 m safety reserve.

[1] This means that a results corridor of the measured dimension of ±1.25 mm around Dimension x would be tolerated, but when the results deviate by more than ±1 mm, correcting the process would be considered.

[2] This reserve is generated externally to the project by the insurer, but would serve as a monetary reserve when the insured event occurs, and the insurer pays.

One may also use a more complex special purpose software, for which a number of products are available. These are all acceptable forms of requirements registers that help capture, document, and track requirements over the course of the project. Again, there are several reasons why it is useful to capture requirements in the form of keywords with some additional information. One reason is the large number of documents in which they can occur. A project may have a Statement of Work (SoW) from the requester or customer, the team may have developed a scope statement, the project manager may have been formally authorized in a project charter, and there may be hundreds of e-mails, or some may have been discussed only in a meeting.

Keeping track of all these requirements can be a difficult task, but project success will definitively depend upon the degree to which they have been met in the end.

Another reason is deliverable acceptance. For a small project "among friends", this may be a quick and informal job, because the team has all the trust that it will resolve problems that later may occur quickly and in an informal and uncomplicated manner. In a larger organization or in a customer project, acceptance will have to be more formalized, and for such cases, a plan for acceptance and handover may be useful. Such a plan can be developed much more easily if there is a requirements register.

In addition, the requirements register can be used during a risk-finding meeting. The group then goes through the register entry by entry and asks, "What risks can be linked to specific requirements?"

5.7 Meetings

Death (of a project) in the Kickoff Meeting. It was late in 1998. A merger project was planned between two groups of companies, Rattlesnake Inc., an organization with over 55,000 employees, and Viper Group, with about 31,000 employees. By that time, both groups had subsidiaries in energy and energy-intensive production industries such as packaging, metals, and chemistry and were fierce competitors in many markets.

The merger project was managed by a consulting firm, Mongoose Ltd., for which I frequently worked as a trainer at that time. The assignment for the merger project gave me a fully booked calendar for the next few months to conduct project management seminars for several hundred sub-project managers from both groups of companies that were going to become one big company. The goal was to link the various Rattlesnake companies with their respective Viper counterparts and identify synergies from mergers, or at least from tight cooperation, between the companies from both groups. This would reduce the number of administrative staff in relation to the productive staff and, at the same time, reduce the pressure on prices and service quality from the competitive situation.

I was considered a core team member of the project and was therefore invited to join the kick-off meeting, during which the project was presented to the top managers of the member companies of both groups. Over 100 top managers from member companies of both groups travelled to one location, and one could hear a nice mix of languages and dialects.

The meeting was opened by the CEOs of both groups. They had some introductory remarks and then introduced the project manager, a senior consultant of the consulting firm Mongoose, who presented the details of the merger project.

During the presentation, a Rattlesnake manager spontaneously stood up, and I remember that he asked a question, "I can of course only speak in the name of my direct colleagues, but I am sure that my concern applies to our Viper colleagues as well. You may not be aware that we are already packed to the rafters with our day-by-day work. We do not have any free capacity to utilize for the project. So, what should we consider more important? The project or our daily business?"

The consultant responded, "The project has of course highest priority. It is the future for all of us. Without the project, our organizations will have no potential, as both are too small for their markets. But . . ." taking a short break, he gazed to the CEOs for confirmation, ". . . but you must at the same time make sure that your daily work will not suffer from the project. Otherwise, our competitors would be laughing up their sleeves."

In this moment, the managers' dissatisfaction was easy to sense.

A reliable contact told me what happened after the meeting. A group of Viper managers called the major shareholder of their company and convinced him to stop the merger, because they regarded it essentially as a takeover. Some weeks later, the failure of the project was officially communicated, of course with some other explanations given.

I had to delete all the entries in my booking calendar and find new business somewhere else. Rattlesnake merged two years later with another group, Boa Corporation, to become a large energy firm; Viper was split into two companies a year later.

What is a kick-off meeting? As a trainer, I observe a common confusion between the term "kick-off meeting" and other common project-related meetings, especially the "on-boarding meeting". On-boarding meetings are performed during the course of the project, when team members or contractors have joined and are introduced to the project and its stakeholders. Rarer, but also observable, is the confusion with the handover meeting, which "kicks off" the use of the deliverables. A project with several handovers can also have several handover meetings. These types of meetings are different and take place at different times, as Figure 5-11 illustrates for a project with five on-boarding meetings and three handover meetings.

A kick-off meeting presents the project to key stakeholders and obtains their formal approval. These stakeholders are required to provide resources to the project, including management attention, people, machines, work space, funding, and more, and their operations or other projects may suffer from disruptions from this project. Normally, a project only has one kick-off meeting but may have multiple on-boarding and handover meetings, because it is normal that team members join the project team at different times and that deliverables are handed over in increments. There are two reasons to plan more than one kick-off meeting. A phased approach may require a kick off the individual phases, so each of them may have a phase kick-off meeting. Another reason may be that the project is a customer project, and it may be appropriate to have two kick-offs: one with the customer to discuss issues that the customer should deal with, and a second, internal kick-off for those issues that are private to the contractor, including the important questions, "How does the project plan to make a profit for the contractor?" and "How will the project be coordinated with the other projects in the contractor's project portfolio?"

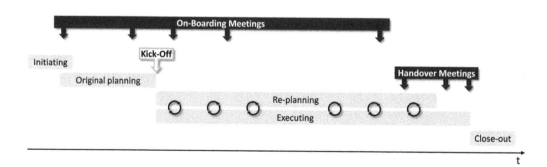

Figure 5-11   A project with several on-boarding and handover meetings but only one kick-off meeting.

An on-boarding meeting may be an excellent opportunity to discuss any team member concerns and objections in a free and open atmosphere with an open agenda. It is often the first meeting in which the team members come together, and it is a good time to discuss assignments and what they mean for the team. A kick-off meeting is different. Attendees are managers, possibly top managers, and any critical questions that cannot be answered satisfactorily may lead to project termination. The kick-off meeting is the opportunity to win management support, but is a time when a project may be terminated, so it needs to be planned and executed in a flawless way.

Who should join a kick-off meeting? At the least, the project manager, supervisors from management groups, and the core team members.

There is no universal definition of what constitutes a core team. I propose a definition that core team members are those stakeholders who are assigned as contributors to the project over the entire life cycle and who will perceive project success as a personal success and project failure as a personal failure.

There are other considerations—for example, operational project stakeholders have to tolerate many inconveniences when they agree to support a project:

- Provision of staff and other resources to the project
- Operational disruptions
- Competition for management attention
- Uncertainty and risk that naturally come with running projects (while operations strive for predictability and order)
- Distress and conflict over staff assignments in matrix organizations
- Additional needs for travel and communications over distances and often over time zones

The kick-off meeting should address these and more issues. It can be a critical moment. Some attendees are high-level managers, and when discussions become scrutinizing, they may finally lead to project termination before its execution has started, as the example showed. Therefore, it takes preparation to conduct a great kick-off meeting and to ensure that the core team can answer any difficult technical questions to assist the project manager.

There is another aspect to consider for kick-off meetings and also for any other moments of communication with management that are intended to bring about decisions: Managers like to base decisions on numbers. Give managers wrong numbers, and they may make poor decisions. Give them no numbers at all, and you may not get any decisions from them.

Handover meetings and acceptance meetings may be conducted during the same event, or they may occur separately. Both are often conducted with a low degree of diligence. The impression in practice is often that the expectation is more to create a list of concerns that will then be used as backlog for the final work rather than as an event that allows the project manager and the team to shine and finish the event with pride and all necessary signatures. After months, possibly years, after the project has been performed, this is not the end that the contributors deserve.

I mentioned the requirements register earlier; it is a useful tool to develop handover and acceptance plans. Simply go through the entries in the requirements register one at a time and decide for each one whether and how they should be verified or validated, transferred, and formally accepted. If a project manager wants to create a checklist for handover and acceptance, the requirements register is the document that helps create it easily and inclusively.

5.8 Scrum

Discussing the dynamics of success and failure, we have already seen concepts that were from sports, such as the distinction of closed-skill and open-skill disciplines discussed in Chapter 1. The word Scrum has been taken over from Rugby, a very open-skilled sport, where it describes a status of interlocking of six or eight players of each team with the "ball" somewhere in the middle. Each team tries to bring the ball under its control by hooking it backwards with the feet to the outside of the scrummage, where a player can take it and carry it away. The term describes a highly competitive moment where both teams' success depends on their combined strength, which is strange, because Scrum as a development method tries not to be competitive at all.

Scrum originated in Japanese product development[*] in the 1980s, and was transferred into software development in the early 1990s by Ken Schwaber, Jeff Sutherland, and others. It was soon carried over to other application areas such as organizational development and engineering. Of the various "Agile" methods, it seems to be the most popular one, and has gained the highest presence in public media and on-line discussions. The early development of Agile methods was focused in the "Agile Manifesto"[†], which was published in 2001, and then implemented in the form of several rule systems including Scrum. The word "manifesto" indicates one of the basic problems with Agile methods in practice: they are sometimes communicated as a kind of ideology.

The reminiscence to the German *Kommunistisches Manifest* of Karl Marx, published in 1848, is obvious. Its contents, however, arise more from the French *Manifeste de l'Anarchie*, published in 1850 by Anselme Bellegarrigue, with his central tenet that *"Oui, l'anarchie c'est l'ordre; car, le gouvernement c'est la guerre civile"*[‡]. A Manifesto is a document that turns a strong personal opinion into a dogma. Common for all manifestos is a black-and-white perception of the world, combined with an "us-and-them" attitude, which either negates the existence of shades of grey or locates them on the "them" side[§]. There is a basic understanding by many of its promoters that Agile principles are indeed a kind of ideology or dogma. Agilism comes not only with self-organization, but also with the presumption that the entire rest of the project world uses harsh command-and-control approaches. This Agilism has made it difficult to assign these methods a valuable place in a tool kit that a project manager should be able to select from and apply in certain situations. When the head of a Project Management Office (PMO) of an international electronics company told me some time ago that it was moving all of its projects to Agile methods, I wondered if it was not replacing one problem with another one. In a situational toolbox, however, which asks for the situational appropriateness of a method or tool before it is used, Agile methods fall into place and should definitely be included; with its popularity and simplicity, Scrum may be one of them.

---

[*] (Sutherland, 2011).

[†] (Beck, K. et al., 2001).

[‡] "Yes, anarchy is order, as government is civil war" (own translation).

[§] In 2007, I had the opportunity to contribute to a German textbook called *Agiles Projektmanagement* by Bernd Österreich and Christian Weiss, with Uwe Vigenshow as another contributor. While my contribution was rather small, I attended the peer review process that the editors conducted with other experts on the matter, and I was stunned to observe a degree of zeal and fanaticism on details. I would rather expect it from competing political or religious groups but not from people in a business environment.

There is a lot of literature on Agile methods on the market, and there is also an abundance of experts available. I will focus here on those aspects of Agile methods, and especially Scrum, that are of interest to include in the toolkit of a situational project manager. The description will not be complete.

The basic idea behind Scrum is to allow late changes to the requirements that a project needs to meet. In my classes, I use an example. I look for the student who lolls most in his chair (I have no idea why, but men tend more to loll in chairs, when women on average prefer to sit upright, at least in my classes) and then ask the person, whom we will call Jack for the exercise: "Jack, are you sitting comfortably?" The person is normally surprised, often changes immediately to an upright position, and answers "Yes, I am". I ask the person, "Great, what does it take for you to sit comfortably?" Jack says, "Well, a comfortable chair". My next question is, "Can you explain what makes a chair comfortable for you?" The answer is, "Of course, I can. It should have four legs (most classroom chairs rather have a set of five swivel casters), it should have arm rests (many classroom chairs do not have that either), a seat cushion in the right height . . ." and so on. Then I ask, "Can you tell us a bit more about the cushion? How soft should it be?" This question makes most students feel slightly uncomfortable. "Well, it should be soft." . . . "Very soft?" . . . "No, not too soft." I go on, "But would you also dislike sitting on hard wood?" "No, I would dislike to sit just on wood. The cushion should be—well—middlesoft."

Then I select another student, Jane, for the example, and talk to the group: "It will surprise you all, but Jane has a secret hobby as she makes custom chairs, perfectly formed for a person's back, and optimized to make the person comfortable. Others tinker on their motorbike in their leisure time, or collect stamps, or knit pullovers, but Jane makes chairs." I turn to Jane: "Would you be able to cushion the chair for Jack to make it 'middlesoft?'" During the following discussion, it becomes clear that "middlesoft" sounds understandable and sufficiently accurate in normal life, but as soon as one has to make a chair cushion "middlesoft", the person will find that the information to do so is insufficient. People who make chairs must have a clear understanding of cushions and how to give them the right softness, but normal people do not have the language, the metrics, or the understanding for accurate statements. This is strange since we sit so many hours a day on chairs—at work, in the car, in mass transport, during meals, when we play with children, and on many other occasions. We have years, possibly decades of experience with chairs, but we are all unable to explain what a proper chair cushion would be. We can do that when we sit on a chair, but in project management, delivering the results to see if they are what the requesters or customers need would probably not be an acceptable approach. I have used this example for over 15 years to describe the principles of Voice of the Customer (VOC), and I didn't have a single student who could answer the question for the perfect seat cushion in sufficient accuracy and detail.

Scrum accepts this basic inability and allows customers and requesters to come late with change requests, which reflect their improved understanding of what they need, based on partially finished results. This is diametrically opposed to traditional principles such as that of old Roman Vitruvius, who demanded over 2,000 years ago that a house should be arranged by three drawings: ground plan, front view, and perspective, which are created through reflection, careful dealing with what is known, and invention—boldly moving forward to discover the unknown. A Scrum project in contrast moves forward by continuous development and delivery in small increments. In Scrum, the way forward is done by walking, and the results that it produces may be different from the requirements and objectives that were known when the project started. This also sets the first limitations on the method: If you have to consider long lead times to book resources and order goods and services; if you have to obtain licenses, approvals, and budgets early in your project; or if you have other requirements that make long-term predictions necessary, Scrum is the wrong method.

There is a set of prerequisites that stem from a principle: Scrum is leaderless. There is no such thing as a "Scrum project manager". This means that a great team with high competence and dedication leads itself, which has already matured through the four Tuckman phases: forming, storming, norming, and performing. This team puts mission first and is prepared to work as a "band of brothers and sisters". Scrum teams are commonly collocated to allow self-organization, and the members are mostly dedicated to the project and almost permanently available, so that the team does not have to deal with availability schedules. An often overlooked aspect of self-organization is that team members must be given time and freedom to coordinate themselves. This nonproductive time is often considered a productivity reserve, when delivery dates become pressing; instead of coordinating itself and its work, the team is then disorganized. The team also needs the freedom to discuss and make decisions, which comes with trust by upper managers and customers in their integrity and professionalism. Most Scrum projects are not decomposed, but composed, so the communication tools and the work environment must support the approach.

Scrum does not define a lifecycle approach, but a fixed set of team roles, artifacts (= documents), meetings and a regular work rhythm (=time boxes). Figure 5-12 shows the basics.

A sprint is a time frame, mostly between one to four weeks, rarely more, during which a working result in the form of a "potentially shippable product" is developed, an increment that could in theory be handed over to users and used. Some will indeed be handed over, as they are the actual releases. The potentially shippable product is developed, tested, and accepted and ready for a theoretical or actual handover. During a sprint, no changes are allowed. The product owner collects the change requests and feeds them into the process in the meantime between two sprints, which is used for the last sprint review and decision making for the next sprint. A second iteration cycle is each development day, which begins with a daily standup meeting to coordinate the team and discuss issues.

Project roles in Scrum are:

- The product owner is the person responsible for the functions and the overall success of the product. The product owner defines and prioritizes the product properties in the

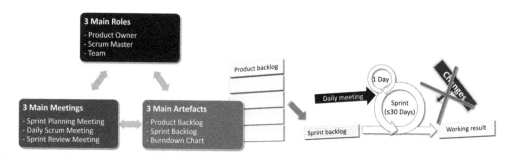

Figure 5-12 The major roles, meetings, and artefacts in a Scrum project, and the sprint iterations based on backlogs.

product backlog and shares responsibility for the economic impact of the product. In a customer project, the product owner represents the customer side of the contract.

- The Scrum master is responsible for the successful implementation of Scrum and removes obstacles. He or she ensures the adherence to the Scrum practices, values, and completeness along with the functioning and the productivity of the team. The Scrum master is understood as a facilitator and method expert rather than as a manager.

- The Scrum team is responsible for the product success and for its incremental delivery. It consists typically of five to nine people from various functions, who work full time for the project.

A Scrum project may have further roles, such as managers, but the roles above are the core roles. Documents used in Scrum are:

- Product backlog, the list of increments (components, functions, and features) of the final product; ideally defined in a way that one could "sell" each of the increments theoretically as a potentially shippable product. The approach is used to decompose work, validate progress, and allow for quick wins.

- Sprint backlog, the list of increments (components, functions, and features) that are expected to be developed during the current sprint.

- Burndown chart, a graphical representation of the remaining tasks and efforts for the current sprint.

- Further artifacts are the impediment backlog, which deals with issues and obstacles, and the "definition of done", which is a checklist of actions that the team must complete, and the results generated by these actions.

Sprint uses three types of meetings:

- Sprint Planning Meeting
- Daily Scrum Meeting
- Sprint Review Meeting

Further meetings are the Release Planning Meeting, the Estimation Meeting, and the Sprint Retrospective Meeting.

A strength of Scrum is the strict discipline regarding the sprint cycles, which balances against the uncertainty on the scope. Scrum requires small, cross-functional, and collocated teams that are assigned full time to the project, have passed the difficult phases of team development, and are dedicated and skilled enough to pass with the customer along a path that needs yet to be explored. It also requires a strong customer/requester representative who dedicates time to the project as product owner and shares responsibility with the team for overall success. This is a combination of requirements, which is quite a good recipe for success in exploratory projects with or without Scrum. Scrum works less well, when the following occurs:

- A high degree of predictiveness is required and predictions are possible.
- A clearly described work flow is necessary early in the project to prevent rework, quality problems, and safety hazards.

- Planning must be granular, and the project manager must enforce it against resistance.
- Proximity of team members to distributed users or markets is more important than between them.
- Clear assignments of responsibilities, but also of costs and effort, are needed.
- Availability of team members is uncertain.
- The sprint duration is insufficient to generate a "potentially shippable product".
- The development work must be done isolated from the requestor or customer.

5.9 PDM Network Diagramming

Agile methods such as Scrum are favorable in exploratory projects, where the way is made by walking. Other projects, however, allow for long-term predictions, many necessitate long-term plans, and Agile methods are insufficient for such projects—they were not developed for them. On these projects, resources are scarce and must be booked early in advance. Agreements with contractors must be made in a timely way, possibly preceded by longer business development phases by both the customer and the contractor. Operations must be able to prepare the working environment in time for the disruptions and changes that the project will bring for them. A workflow plan is often necessary to understand and visualize how activities interrelate and to validate timely delivery or show the impossibility of meeting deadlines at an early stage. The predictability of operations, their most critical business asset, is often heavily impacted by the disappointments that almost inevitably result from poorly planned and managed projects. These disappointments can impact the organizations involved or entire societies. Following a "we resolve problems as they occur" strategy is a recipe for failure when the moment of occurrence of a problem comes too late, and when all options to effectively manage it have already vanished.

Predictive approaches may include many knowledge areas. The project team may have a cost plan based not only on price lists of suppliers, but also on forecasts of billable or internally chargeable work or availability times. Workloads must be predicted to ensure that resources are booked in sufficient quantities over appropriate periods. We may have various management plans that describe our project management processes, responsibilities, methods, tools, etc. that we are going to use in the project, and so on. My focus in the following paragraphs will again be on the time aspect, because this is the most complex one. As already mentioned, costs are comparatively easy to calculate since the sum of all individual costs in the project is the total cost. If the cost of a work item increases by $1,000, the costs of the entire project increase by the same amount, and money saved in one work item can easily be used to cover additional costs in another one. Money is universal. Managing work amount or effort is slightly more difficult. The sum of all work amounts is the total work in the project, and if the work amount in an activity increases by 10 person-days, the total work of the project will increase by the same amount. Work is more complicated than cost because a resource that unexpectedly became available in one activity may not be able to do unexpected additional work in another one. Humans may have the wrong qualifications, machines the wrong functionality or power, and the geographic distance between the locations of the two activities may be another obstacle. People's egos should also not be under-estimated—you cannot require a developer to clean the restrooms.

Time is different. Adding up individual durations of work items does not give us the total durations. Some activities can be done concurrently, and one should never underestimate the effect of idle times—mostly neglected in literature and education but a strong impact factor in reality—which some teams may be able to reduce to a minimum but will never be able to avoid completely. Instead of simply adding up individual numbers to obtain the total numbers, as we would do with work and costs, we need to use a more sophisticated tool, which is network diagramming. Chapter 4 discussed the history of four concurrent developments in the late 1950s and early 1960s in the USA and France, called PERT, CPM, PDM, and MPM. The survivor of this competition was PDM with elements of CPM and MPM included and further enhanced by special software, which upheld the underlying mathematics but changed the standard graphical representation to what is now commonly, but falsely*, referred to a Gantt chart. In addition, software adds restrictions of resource availability as another element to network planning, which allows the creation of far more realistic plans. Resource bottlenecks influence project progress much more than what was accounted for in the original methods. Network diagramming can help predict the future up to a long planning horizon, depending on the completeness, the realism, and the reliability of the basic data and the work flow model used.

Figure 5-13 shows an example of a simple network diagram in PDM notation, which means that activities are drawn as nodes (boxes), dependencies between them as arrows.

The diagram shown follows some basic rules and simplifications that make the example easy to understand:

---

* (Weaver, 2012).

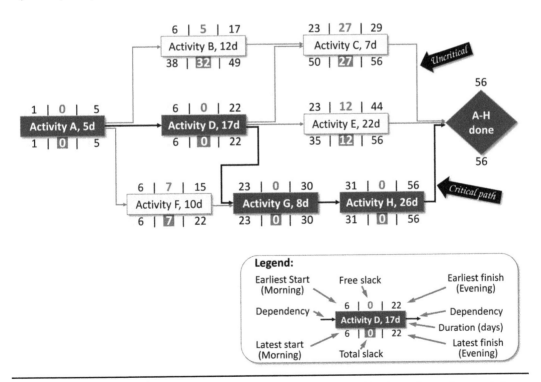

Figure 5-13   A simple network diagram with forward pass (top) and backward pass (bottom) dates.

- It is assumed that activities take full days to be finished.
- Days are shown as working days. If the first day is a Monday, the sixth day will be a Monday too.
- Holidays and other interruptions will not occur.
- It is assumed that a working day starts in the morning and ends in the evening.
- In the calculations of slack, it is also assumed that all activities are started at their earliest dates.

It is assumed that no resource shortages will impact the schedule. When the network, for instance, allows Activity D and Activity F to be performed in parallel, there are enough resources available for that. It is also assumed that resources are always available at any time, when they are needed.

I discussed earlier that this last point is probably the weakest point of many planning methods, including Agile and predictive methods: the underlying presumption that resources are generally available when they are needed. My experience and observations are rather that resources are not available when they are needed and that the project manager must work hard to acquire them. They may not be available full time, and promised availability may later be discarded when the project has to compete with more urgent undertakings of the performing organization. Managing both a work flow and the often volatile availability of resources manually is a task that can overpower a project manager, hence my recommendation to use project management software, which can ease this job significantly in the hands of a well-trained person and allows for new predictions when they are needed. In this chapter, the discussion will focus on predictable projects and predictive approaches; we will therefore ignore this question for the following discussion.

When the diagram has been developed based on the workflow identified or designed by the project manager and the team, and when duration estimates have been made (I recommend PERT 3-point estimation), a forward pass series of calculations has been made to show the earliest start and end dates for the activities and identify the earliest date for the milestone, simply by going left to right through the diagram and adding up the durations of the activities. They have been denoted in the top left and top right corners of the nodes. Then, a backward pass series of calculations has been made from the milestone date going left, subtracting the durations. This gives us the latest dates for the activities, denoted in the bottom left and bottom right corners.

For the forward pass calculations, we have to take care of sink nodes, also called convergences, the points at which parallel paths merge. The beginning of Activities C and G are sinks, and this is also true for the milestone. At a sink, the successor can only be calculated when the predecessor has been calculated. During the backward pass calculations, we have to consider burst nodes, also called divergences, places at which a diagram path splits in two. In the example, this happens at the end of Activities A and D. The latest dates for burst nodes can only be calculated when the dates for the successors have been planned.

The diagram also shows free and total slack[*]. Free slack is the number of days that an activity can be delayed without affecting another activity. Total slack is the number of days that an activity can be delayed without impacting the earliest milestone date (as shown in Figure 5-14).

---

[*] Sometimes called "float" or "play".

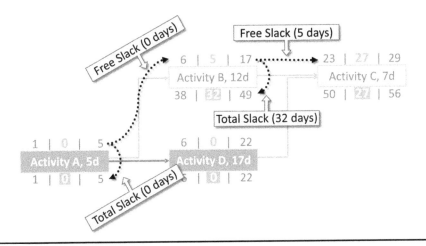

Figure 5-14   Free slack and total slack in a sample diagram.

The approach makes it easy to identify the critical workflow in the diagram, commonly called the critical path: the work stream with total slacks = 0, or in other words, identical forward pass and backward pass dates. This is appropriate for projects that should be finished as soon as possible, or have as much time as needed to get all the work done. How many projects enjoy such luxury?

In the informal survey as of September 2015 that is discussed in Chapter 3, 23.7% of the respondents responded that this is their reality. Over 75% said that they have to perform their project against one or more deadlines. This is reflected in Figure 5-15, in which the backward calculation begins at a time constraint, a deadline.

With this form of backward pass calculation, the total slack on the critical path is not zero. Instead, the critical path now is the flow of activities with the least total slack. This version has

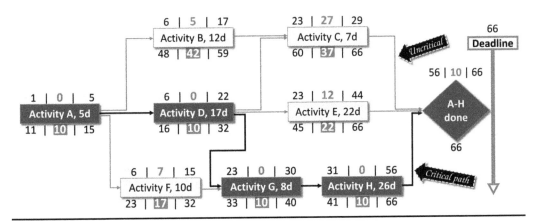

Figure 5-15   The same simple network diagram with the backward pass (bottom numbers) calculated from a deadline.

the benefit that it shows us for what time we need to book resources. If Jill is expected to do Activity D for us, she will not be able to start before the morning of the sixth day, and if she will not be able to start until the morning of the 16th day, we will miss the deadline (presuming, of course, that the estimates for activity durations and for workflow dependencies are correct). Problems with the timeliness of work become visible weeks, often months at a time, when there are still many options for decision making, and when these options are still relative inexpensive. Using network diagramming, we powerfully skew the learning curve as Chapter 2 describes. Our ability to meet deadlines increases while project costs are likely to decrease.

When network diagramming was invented, the alternatives were to use PDM or MDM for manual use on walls or blackboards or the much simpler methods CPM and PERT that were developed for use with mainframe computers that needed extensive programming and data entry. Today, this all has been simplified by using software programs, which range from free and simple solutions with limited functionality, to standard programs at moderate costs, to highly sophisticated solutions that support complex simulations and cost a five-digit dollar amount. The offerings are numerous, and in addition, one can chose between desktop programs or Software as a Service (SaaS) solutions provided over the Internet. Figure 5-16 shows the example modeled in software. The days are now not counted as workdays but as elapsed time, which allows the ability to plan for holidays and other restrictions of availability and does not force us to plan with full days.

The early information available enables the project manager to book resources and contractors and at periods that are compatible with imposed deadlines. This allows the project manager to increase the chance to deliver on time, which, in my survey, was necessary for the vast majority of projects.

The previous discussion describes the basics of network diagramming. There are many more things to learn on this method, such as the different types of dependencies, the leveling of plans to account for limited resources, or the expansion of the CPM to become Critical Chain, which does activities as late as possible, adding time buffers to avoid finishing them too late. However, this information is beyond the scope of this book, and there is a great amount of literature available for those who wish to learn more.

Network diagramming as a tool for predictive project management has many benefits. It allows for long-term predictions and gives indications when internal resources need to be booked and contracts for external resources need to be closed and when they need to be made available by the contractor. Planning the flow of work visualizes dependencies among activities

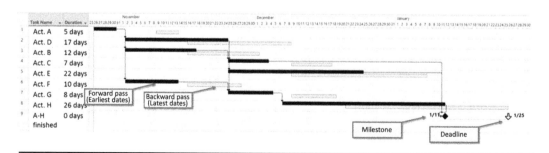

Figure 5-16 The same network diagram drawn with software. The earliest dates are shown as black bars, the latest dates calculated from the deadline are grey.

whose ignorance can otherwise lead to massive amounts of rework and to quality and security problems. In addition, operational disruptions can be agreed upon and planned, which benefits affected functional managers. In production, for example, Line A, whose products are used as inputs in Line B, must be closed for some weeks. Timely planning enables you to develop some stocks such that Line B does not run out of inputs. This approach enables the organization to go on with its production, rather than resulting in the production stop of Line A causing a production stop for Line B, and possibly many others too.

Problems with predictive approaches can occur when deviations take place. For example, an activity takes much longer than planned, and the booking times and deliveries ordered need to be rescheduled. Even more difficult to manage are change requests, which may necessitate adding further activities or changing the order of existing ones. The hardest problem is often when stakeholders cannot describe their requirements, which makes it impossible to define the project scope and develop activities against it. Another problem can be surprises, which let the team experience the power of the unpredictable.

5.10 Situational Project Scheduling

Modern Rome is the capital of Italy, a city with a population of 2.6 million. Chicago in the USA is comparable in size (2.7 million), and so is Osaka in Japan (2.67 million). Chicago has an urban rapid transit system with eight lines operating over a network with a total length of 165 km (103 miles). Osaka has a combination of publicly and privately held transit networks with a total of 70 lines. One of these networks is the Osaka Municipal Subway, which has nine lines and a system length of 138 km (86 miles). Rome's Metro, in comparison, is small: three lines serve a network of 60 km (37 miles). It is not yet even a network because the newest line is still under construction and is not even connected with the other two lines[*].

Projects to build urban rapid transit systems under thriving city operations is a difficult task in all cities around the world, but Rome has the special difficulty of the ancient history that lies buried under the modern town. When the tunnel cutters under the city find ancient buildings or catacombs, they have to stop their work, possibly for months, to allow archaeologists to rescue as much as possible from inside the rooms before the project can continue. Rome's tunneling engineers made a decision at one point to go deeper, under the level of the ancient city, but they have to build access ways to stations and venting systems that connect the subway with the surface and need to dig through the ancient levels.

The modern Roman construction engineers must follow a mixed approach. The dynamics of success and failure dictate for them that they do long-term predictions and plans to have resources available when they are needed, but they must also be prepared for changing the plan when the need emerges. They have to balance their path between predictive and Agile and must be prepared to quickly react to changing environments, moving their practices at times more to the predictive or at others to the Agile side, depending on the needs and the opportunities of the situation. They plan ahead as far into the future as they safely can do, and they have a less detailed plan for a longer term, which will be refined when things go as planned or will be re-planned if they do not. This is probably the situation that most project

---

[*] In October 2015.

	Agile		Waterfall
Approach	Exploratory		Predictive
Specifications, requirements	Often changing or unknown		Clear, detailed and static
Environment	Fast changing		Rather static
Change requests welcome	Mostly		Rarely
Emphasis on	Quick wins		Long-term goals
Work structured by	Product backlog		WBS
Planning cycles	Sprint and release planning		Long-term network planning
Planning granularity	High-level planning		Granular planning
Team structures	Self-organization		Command & control
Preferred software	Collaboration tools		Planning and tracking tools
Preferred team size	Small teams		Any team size
Preferred team mode	Collocation		Collocation and virtual
Preferred delivery mode	Staged or continuous	Situational	Single or staged

Figure 5-17 SitPM is open to lean more to a predictive or Agile approach, depending on the needs and opportunities of a situation.

managers find themselves in, according to 58% of the respondents in my ad hoc survey in 2012, and the rolling wave approach is definitively an appropriate one, never a perfect solution but allowing to make the best from ever-changing conditions under which the project must be performed.

We have to also take care of contracting. Bonus/malus clauses using penalties, liquidated damages, or incentive fees rarely make allowances for adaptiveness, and with every change that is decided upon, the effect on contracts should be considered. We also have to take care of the people involved, since too many changes in a short time may unsettle and frustrate those among them who expect projects to be steady and predictable. Another benefit when using a rolling wave approach, also known as progressive elaboration or iterative incremental, is that it leaves no room for ideology. Located in the middle between predictive and Agile approach (see Figure 5-17), the project manager and the team can at times lean more to the Agile side, at others more to the predictive—a pragmatic flexibility that dogmatic Agilists and Predictionists rarely have. It may also be appropriate that parallel work streams are performed with different planning horizons, levels of granularity in planning and tracking, and different degrees of openness for changes.

Using this open approach, both Agile and predictive planning can be situational tools.

5.11 Staged Response Diagram (SRD)

Here is one more tool that I have used for some time and seen implemented in projects with very positive results. I called it Staged Response Diagram, or SRD. It is a combination of two methods:

- Meilenstein-Trendanalyse, Milestone Trend Analysis (MTA), a popular tool in German project management.
- Limit definition, which is popular in international quality management.

Picture a situation in which a flow of activities has to end with a milestone, which is linked to the delivery of software, a machine, or any other product. The deadline for the delivery is June 30, 2016. To protect the deadline, the project manager defined two more dates as limits:

- A Control Limit on June 10, 2016, 20 days before the deadline. If the projected delivery date is forecast to be later than this date, corrective action will be considered to accelerate the project.
- A Warning Limit on May 31, 2016, 30 days before the deadline. As long as the delivery date is projected for a date earlier than this limit, the team remains relaxed and focused just on its work. If the limit is exceeded, attention is increased: The frequency of reports and meetings may increase, inexpensive measuring equipment is replaced with more accurate and more costly items, the project manager tries to get routine work delegated to have more time for the project, and so on.

Another prerequisite is the use of network diagramming to replace guesses with valid projections based on estimates and actuals. Software can help by accelerating and easing the calculation and drawing process. The project manager then puts the data in a diagram as shown in Figure 5-18 to observe the behavior of the diagram against the deadline and the limits. The

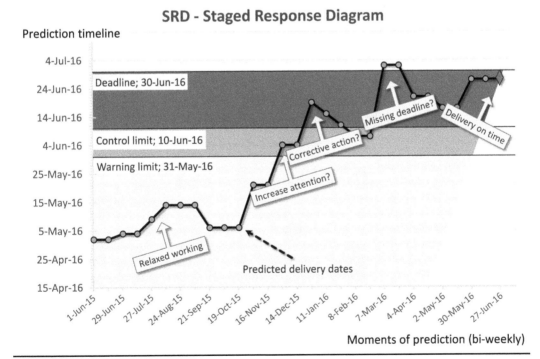

Figure 5-18 An SRD with limits that influence the behavior of the project manager and the team to ensure timely delivery.

X-axis in the diagram represents the dates when the projections are made. The Y-axis is used for the projected delivery date. The SRD tool raises awareness and triggers actions when this is appropriate and documents also when this has happened and why.

The core benefit of this tool is that it endorses situational behavior and documents a learning process over time.

I am often asked by students how frequently a team should have meetings, and how much time the team should spend on them. A similar question relates to the frequency and level of detail for reports. I do not have a universal answer to these questions. Too much time spent in meetings and writing reports will inevitably disrupt the development process. Not enough time may lead to developing the wrong things, and it may become too late for the project manager and the team to become aware of this problem. Then, options for reaction will be scarce and expensive, which we already know as the common effect of late responses. Communication and attention are generally expensive. Corrective actions are also not free and should be reserved for the times when they are needed. The multi-step approach in the diagram makes sure the attention to the project matches its risk exposure, which grows with the proximity of the projected delivery date and the deadline, and corrective action is not applied out of nervousness at the wrong time, but also not missed, when it is due. The same tool can of course be used for cost predictions against budgets and funding limitations, for workloads, times of operational disruption, and other aspects, where project managers must adhere to constraints.

5.12 The Stakeholder Attitudes Influence Chart

Stakeholder management has been described in various sections above. With its various elements, including stakeholder identification, analysis, expectations setting, and engagement,

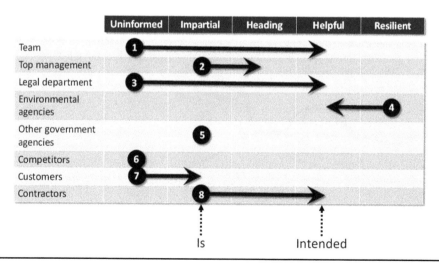

Figure 5-19   SAIC (Stakeholder Attitudes Influence Chart). Numbers represent relative prioritization of the stakeholders.

it is a difficult task which requires skills for which most project managers have not had much if any training, at least not in a formal sense. It is a key task for projects that have a strong impact on people, groups, communities, and organizations, intending to reduce resistance and strengthen support and engagement. It is also deeply situational, as different people, groups, business units, organizations, and so on may have different requirements on the project, and these requirements are likely to change over the course of the project.

Stakeholder management can put a high demand on the project regarding time and cost. This raises the need for prioritization, which the Stakeholder Attitudes Influence Chart (SAIC) can support. The SAIC helps manage and prioritize stakeholder-related activities. In the example in Figure 5-19, the project team members have been given the highest priority (lowest number) and the contractors have lowest priority.

5.13 Turturism, Private Settings, and Leadership

Michael Ende is the author of children's books, mostly known for the novel *Never-Ending Story*, which became the basis for a sequel of three movies between 1984 and 1994. Internationally less well-known are his novel *Jim Button and Luke the Engine Driver*. In this story, Michael Ende portrays a strange Mr. Turtur, a virtual giant, who looks frighteningly huge from a distance, but instead of getting even taller when one approaches him, he paradoxically shrinks, and once one is next to him, he attains the perfectly normal size of an average person. Mr. Turtur spends a lonely life, because people are afraid and stay away from him, while he is actually a friendly and helpful person.

In project management, we often have to deal with virtual giants, persons who seem to be strong, powerful, and frightening from a distance but appear just as normal humans in close proximity. They mandate projects as tough sponsors or build up resistance as opposing stakeholders who reject the impact of the project on their interests. Turturism may seem an illogical effect, but it is often present in project management.

On the other hand, the most vicious source of pressure can often be relaxed when we overcome the official interpersonal, organizational, and physical distance and meet in a more comfortable and private setting. Outside the formal business environment of the office, in the meeting rooms and coffee zones inside the corporation, it may be easier for both the project manager and the virtual giant to overcome deadlocks and conflicts of interests and develop mutual empathy. It may sound surprising, but suggesting an appointment with a difficult key stakeholder in a non-business location is a tool for situational project management, as it helps both parties identify commonalities and joint interests. One should know what kind of private settings this Mr. or Ms. Turtur prefers and where he or she would feel most comfortable while reducing to normal human size. It could be a golf or tennis match, an invitation to a football or baseball game, or simply a dinner. Critical decisions are often better made in such a private setting or shortly afterward, when the feeling of commonality and empathy is still alive. It may be a good idea to reserve a small part of the project budget for such events, small enough to avoid the impression of corruption, but sufficient to overcome the distance that makes virtual giants seem taller than they are.

There is a tight connection between empathy and leadership, which is often overlooked. Managers often believe that taking an authoritarian and tough position toward people is a sign of strength and leadership, overlooking the fact that a good leader often better steps to the side, sensing when a qualified team is happy to resolve all problems and attain the critical objectives of the managed organization as long as no one stands in their way, obstructing them and causing frustration.

In the next chapter, I will discuss a view on leadership that is different from the top-down understanding of leadership that some people consider outdated, but that is still common among managers.

Chapter 6

Leadership and the Dynamics of Success and Failure

The word "leader" has some strange specifics: It has a positive connotation in all languages that I am aware of, except two languages, German and Italian. *"Führer"* and *"Duce"* are words that one would not use in the two countries with enthusiasm, rather with great care to avoid misunderstandings, with repugnant reference to history, or not at all. Sometimes they are used as an insult. Whenever I discuss leadership in classes in Germany, Austria, or Switzerland, I have to be careful about the words that I am choosing. George Reed wrote a brief article on the continuum between great and toxic leaders (2014)*. He correctly describes how this leadership continuum can be found in the military, where he comes from. It exists everywhere humans organize to generate outcomes, so it is obviously a topic also for project management.

Military research has an interesting discipline, which deals with historic battles and strategic warfare. Battles are often analyzed to teach students the approaches and stratagems that were successfully applied and won the battle. Often environmental influences such as landscape or weather, and sometimes also psychology and heroism, are also discussed. Much less emphasis is given to the failures and the foolishness that made the leader of the other party lose—failures whose basic principles can be observed as well in any management discipline, including project management. These are failures to meet goals and serve the country, but also failures of responsibility; the losing leaders were in charge of thousands of men, each of them a son, a husband, or a father, and they sent them into a traumatic experience that killed many of them and left many survivors mutilated in body and soul.

One of the most famous fights was the Battle of Agincourt in 1415, which Shakespeare used to give credit to Henry V of England as a great leader who won the battle with a war-weary and lightly armored army against a highly organized, heavily armored, and numerically superior French army (which seems historically correct). The battle took place on a freshly plowed field, soaked with water from a recent rain, and the place was narrow. Contemporaries describe how

---

* (Reed, 2014).

the heavy armor made the French knights sink in the soft ground to their knees and deeper. When they arrived at the combat zone, they were already fatigued, and while the first dead bodies lay on the ground, the following knights stumbled over them, pushed forward by those behind them. The narrowness of the location forced the French to stand densely packed, which made them unable to make full use of their weapons and also made them a perfect target for the English long-bow archers, who only had to send their arrows into the compact mass of men to make their hits and thin out the French ranks. The descriptions by contemporaries of the condition of the French army remember more of the modern mass panics than of combat zones, and it is not hard to imagine the horror that must have taken grip of the men in the field when they saw other men around them dying, and they were able neither to move forward and engage the enemy nor to run away. There are also reports that French soldiers standing in rear ranks ready for their deployment refused to march onto the soaked battlefield, which seems plausible considering what they saw happening just ahead of them. Estimates of the number of soldiers killed on the French side are between 4,000 and 11,000, while the numbers on the English side are given as between 100 and 1,600.

Historians tend to blame the conditions on the battleground for the defeat, often ignoring a very basic question: How could the French commanders send their soldiers onto the muddy field? The battle of Agincourt is to some degree an example of a victorious strategy of a king and on the influence of weather on the outcomes of battles. It is even more a story of incredible failure, caused by non-situational leaders who trusted in the superiority of their configuration, either uninformed of what was lying ahead of them or being aware of the risks but ignoring them out of arrogance and self-centeredness. They had their best practices and followed them blindly. At Agincourt, one thoughtless decision was sufficient to destroy one of the most powerful armies of its time and change the course of history. And this brings us back to project management, in which we also see both outstanding successes and unbelievable fiascos. In both cases, luck plays a role, but often the outcomes are the result of the varying situational qualities of leaders and whether they respond intelligently to the conditions surrounding the project. Success is more than not failing, but avoiding failure is a good principle to build successes on.

To conclude this exploratory journey into the dynamics of success and failure and the recommendation of a more situational approach to project management, I would like to look at the other important element on top of the technical, organizational, and interpersonal skills that a project managers are expected to master: leadership.

6.1 So, What Is Leadership?

Some years ago I asked Professor Jean Lipman-Blumen, the Grande Dame of leadership research, whose work very much influenced Chapters 3 and 4 of this book, for her definition of the word, and her answer was, "I cannot define the word leadership, but when I meet a leader, I make that out immediately". Her former mentor, leadership guru Peter F. Drucker*, considered "thinking through the organization's mission, defining it, and establishing it, clearly and visibly" to be

---

* J. Lipman-Blumen quotes Peter F. Drucker, saying that the word "guru" is often used by newspapers, because the word "charlatan" would be too long for their headlines.

the foundation of effective leadership*. He continues that a leader must see leadership not as rank and privilege but as responsibility, and that an effective leader considers creating human energies and human vision the ultimate task. For Drucker, earning trust is the final requirement, because without trust, there are no followers. While these are all valid descriptions of tasks that leaders fulfill, they are not a definition of what leadership actually is.

Other sources define leadership as "the ability to get things done through voluntary contribution by others". I am not happy with the definition. It tries to define leadership by its effects, when a definition should help us understand its essence. Here is my opinion, based on experience and observations:

It was August 11, 1999, when a total eclipse of the sun was announced for major parts of Central Europe, including Münich, Bavaria, where I had a classroom training that day. The eclipse was to happen about an hour after lunchtime, so I made a decision with my students to stay longer after lunch break and watch the phenomenon together.

At the time of the eclipse, there were thick dark clouds hanging deep, and most people in Münich had no chance to actually watch it. All they saw was the world around them slowly getting dark, an evening just after lunch time, immediately followed by something similar to a morning. My students and I were lucky, however. We happened to find a major hole in the clouds above the restaurant that the training company had booked for our lunchtime, big and stationary enough to see the eclipse from start almost to the end. So, we stayed a bit longer at our lunchtime place. We sat on the terrace of the Michaeligarten park restaurant, which has a nice vista over a little lake and a free sightline into the sky. Just on time, as forecast, the sun circle became dented, and this dent—actually the moon covering parts of the sun—became wider and wider.

On that day, some flocks of greylag geese flew over the park from west to east, and when the moon covered around 90% of the sun's area, a flock of about 40 birds landed on the water of the lake. As geese generally do, they had followed a leader goose, probably on the way to a meadow where they could graze, and this individual must have gotten confused by the slowly progressing disappearance of the sun. After the leader goose had landed on the lake, the other geese followed.

Swimming on the water, the geese were very noisy. The puzzling of their leader unsettled the other geese, and the impression was that they were frightened and did not know what to do. I had not known before how much noise a flock of 40 scared geese could make. Approximately one raucous minute later, another group of geese flew over the lake. This flock was more than twice as large, and they flew in the same eastern direction as the first groups. Their leader goose did not seem confused at all by the vanishing sun but instead flew ahead on its way, and so did its followers in the graceful V-shape order as geese normally do.

The frightened geese down on the lake reacted immediately. They started from the water, which made even more clamor when 40 pairs of wings flapped on the water, left the lake, and each goose found its place behind the wing tip of another goose, a place which gives them the comfort of being securely guided and of reduced air drag. They saw someone who seemed to know the way, and they preferred to follow the leader, which reduced their distress considerably.

I am not sure if anyone else in the Michaeligarten had observed this short episode of less than two minutes. People around me had their protection goggles on and were looking into the sky. I also did that, of course, when this short intermezzo was over.

---

* (Drucker, 2001, p. 270).

But the event remained in my mind, and over the years, I reflected on it more and more. Hasn't 1973 Nobel Prize winner Austrian scientist Konrad Lorenz taught us to use geese behavior as a model for certain facets of human behavior? Humans are not geese, of course, but the dynamics between leaders and followers are probably quite common among different species. I concluded that an individual slips into the role not because he or she has been trained in specific leadership skills or has been placed in a powerful position, but because others have the impression that he or she knows the way, and that it is a good idea to follow that leader. The impression is often enforced or even caused by others, who already follow the leader. This signal may be wrong. It may be a factoid, but factoids can be powerful. Leadership in this understanding is more an expression of group behavior of the followers than of the personal skills or traits of the leading person. It also shows the difference between an Alpha animal, which gets into the leading position by winning a series of power play-offs against others, and a leader in a more narrow sense, who is in the position based on consensus of the followers. The consensus may be a result of swarm intelligence or swarm imbecility. Germany under Adolf Hitler between 1933 and 1945 is a good example of the second, and so are the other evil dictators of the 20th century, and all others before and after them. The impression that someone knows the way may be strong, but the followers may overlook how much the way is paved with blood.

Is leadership as a position arranged more by the followers than the leaders? What about the effect on the leading individual? The followers are perceived as a confirmation for this person that he or she is doing something better than others. It is impossible to look into the mind of a goose, but it is imaginable that the leader of the second flock felt so much empowered by the large group behind him or her that the vanishing sun came much less as something unsettling or threatening. The leader-goose's self-esteem and self-confidence, or the feeling of being in charge of the geese behind, were strong enough to manage the unusual situation. The confirmation that the individual in the leadership role gets will enforce some behaviors more—those which the followers reward—and will weaken others that the followership ignores or even rejects. An additional imprinting effect may be the relationship with other leaders, which may be competitive, collaborative, or even true friendship, and also experiences of gains and losses during childhood. Over the years, this directs leaders to express the achieving styles profiles that were a topic in Chapter 3 and whose application in project management was discussed in Chapter 4.

Here is my definition:

Leadership is the authorization to lead, given voluntarily by followers.

Without such authorization, many self-assigned leaders are either just bureaucrats in powerful formal positions or authoritarian oppressors. The authorization may be implicit by actions or explicit by elections, and the followers may be lifetime devotees who follow the leader tightly and absorb every word the leader speaks, or they may be those who have made a more rational, distant, and temporary decision to follow the leader—something one would expect from members of most project teams. This definition reverses the causality in the dynamics of success and failure of leadership. Leaders become a far more passive part of the relationship than normally assumed: Their actions are not what makes them leaders, but they help followers to decide whether they want them to lead them or not. According to this definition, the true driving force in leadership are the followers. Strong leadership includes being a role model for the followers, but in this definition, the followers assign the role model function to the leader, rather than the leader telling the followers: "Follow my example".

Are leaders then just passive front runners, driven by the followers and possibly by some extraordinary fate into their role? Of course not. They are required to balance the respect that their followers expect and in their understanding deserve, the positive confirmation of reward for efforts endowed by their followers, but also careful corrective feedback when necessary. One of the greatest problem that leaderless organizations (and projects) face is the lack of an accepted leader who praises employees and team members when they have done an outstanding job and celebrates their achievements in front of others.

For project managers, standing still at this point of passivity would not be sufficient. Project managers are expected to be active drivers and agents of change; they are administrators of investments, and when we have been untrusted with the management of these changes, we have to develop impact on ourselves and on others. There is also rarely sufficient time to let the leadership role grow; we must get our teams up to productivity in the short term. The leadership role has some confirmation for us and may be flattering, but it is also uncomfortable, as it comes with expectations on our behavioral self-control—something we may first have to develop—and on additional energy that we can call upon when required. All geese in the V-formation have the comfort of the reduced drag—except the leader*.

When a requirement is made to become a "well-rounded project manager", this definitively does not relate to the shape of the body but to the behavioral traits of the person. In Chapter 2, I discussed how project managers learn, and there is an element that was still missing in the analysis: Along the projects that we lead, we also learn a lot about ourselves. The ever-changing situations that we face teach us lessons on our strengths and weaknesses on technical, organizational, and interpersonal levels and on the preparedness of people to follow us on the way that we want them to go with us. The feedback that we get can be very direct: Technical failure, over- or under-organization, misunderstandings, conflicts, and team attrition are often early warning signs that the way we approach a situation and the character of this situation may not match. Sometimes, the feedback takes longer—when Nash Equilibria are growing almost invisibly until they get the power to wreck our project, or when time bombs are placed in concealed locations, remaining inactive for a long time until they suddenly go off. We also receive positive feedback through successes and the wows of impressed stakeholders. When these experiences tell us something about our own shortcomings, we will need to react. We may enhance the underdeveloped behaviors through training, possibly self-training, to better be able master situations that we experience as especially difficult and challenging, or we accept our shortcomings and avoid situations, where possible, that we feel unable to manage.

In all these achieving situations that we face, we will often have to protect the control that we have over our own professional destiny and that of our projects in collaborative environments. In SitPM, we achieve this protection best by upholding our most profound personal assets: Empathy with stakeholders, authenticity, trustworthiness, integrity, situational intelligence, and by knowing the team and the other stakeholders.

---

* It has been reported that on long-distance flights, geese change the lead goose when the flight work gets too exhausting for the leader.

6.2 As Project Leaders, What Should We Do?

I mentioned Jean Lipman-Blumen* in the discussion of Achieving Styles in Chapter 4. She also describes a tension field, in which leaders act between two extremes, which she calls "Diversity" and "Interdependence". In project management language, we respond to this tension field through "Heroism" and "Collaboration".

Heroes and heroines fight alone. They are much stronger, faster, wittier, or cleverer than others and resolve problems singlehandedly, and they are prepared to walk alone the ways that lead to the achievement of their goals. The role of their followers may as feeble observers or as recipients of commands and direction. Leadership for heroes and followers is more a kind of closed-skill discipline, in which they develop and perform their programs in as much isolation, independence and possibly even opposition to others and the environment of their leadership domain as possible. If they consider it necessary, they build walls, physical as much as mental, to shelter their internal processes from uncertainty and external change.

On the other end of this continuum is collaboration. Collaborative leaders rely on others, and these others can in turn rely on them. Understanding the needs in interdependent societies for constructive mutuality, they try to achieve their goals by building alliances, thus combining their assets and resources with those of others. While heroes build walls and trenches, connective leaders permeate the walls with gates and build bridges over the trenches.

Lipman-Blumen describes three eras of leadership that are dominated by different behaviors of leaders in this tension field:

- **The Physical Era.** This era began in prehistoric times. The dominions of leaders were limited by natural borders such as mountains, major rivers, or simply vast distances over land and sea, and the leaders had to give protection to their followers inside these dominions, both from the dangers that nature brought about and those that came from time to time from marauding invaders or internally competing leaders.

- **The Geopolitical Era.** Leaders defined boundaries—not only in tangible form around their countries but also in the form of religions or ideologies—with which they separated their dominions from those of other leaders. When they developed alliances, their purpose was more to strengthen the borders against their enemies than to actually build strong bridges. NATO and the Warsaw Pact were examples in the 20th century, and they were both kept together for decades over the real or assumed need on each side to protect the "Iron Curtain" against attacks from the other alliance.

- **The Connective Era.** In this era, the connections among concepts, people, and the environment are tightening. People seek to roam much more freely not only through geographical spaces, but also through cultures and understandings, and physical borders no longer protect cultures or powers from external influences. Alliances are forged much less to protect borders than to address emerging global problems, such as environmental hazards or diseases that have the potential to turn into world-wide epidemics. Lipman-Blumen describes the role of technology in this development. In the years since she published her theory in 1996, the connective aspect of technology has become even stronger with the success of the Internet, wireless mobile systems, and social networks.

---

* (Lipman-Blumen, 1996, pp. 4–11).

One should also not underestimate the influence of the deep-sea container in simplifying world trade. Business relationships have been internationalized with an unprecedented intensity, and when "globalization" became a political catchword around the change of the millennium—as a label of new hope by some, but of threats by others—it has simply become an accepted fact since that time, with all the positive and negative implications that came with it.

While Lipman-Blumen developed her analyses mostly from a political perspective, applied in project management, the model brings insights as well. Navigating between the monsters of too much heroism on the one hand and getting lost in collaborative groups on the other is among the most difficult leadership tasks for project managers. Positive heroism consists for them in being role models for their stakeholders, in developing a sense of urgency without pressure, and in maintaining the focus on the project's objectives and deliverables.

But there is a flipside to this view of practiced autonomy: The need to rely on others and embed the projects in an environment that includes the performing organization, an "internal" or paying customer, contractors, and many other stakeholders from outside these organizations. The organizations involved present many chances and opportunities not only for support, satisfaction, and success to project managers and their teams, but also for perceptions of abandonment, frustration, and failure.

To make things more difficult, many project managers perceive themselves as odd men or women out inside their organizations. Unless the project is performed for a paying customer, the core processes of the organization are more likely to be focused on and optimized for operations—the business domain that provides the income to many organizations and gives them the justification for their existence. Project managers in customer projects are often much stronger in their own organization—where they provide the income and the business rationale—but they must spend a lot of time with the stakeholders on the customer side, where they in turn tend to be the misfits because they come from project management as well as from an external company.

In all these organizations, project managers must make resources free for their projects, get access to running systems to obtain information or implement changes, and spend money not only as obvious and visible costs, but also as the hidden costs of operational disruptions. While they are facing a lot of resistance, they have to find ways to engage adversarial stakeholders and redirect their energy from their opposition to a more productive and constructive path.

For this kind of challenge, there are no universal tools that a book can offer to project managers. The task is again fundamentally situational, and the technique that was applied successfully to master one situation may fail in another one. The recommendations that a trainer and book author can give to project managers in these situations are mostly "master your own frustrations" and "know your stakeholders and adjust your approaches to them". The first is necessary, because grudges and anger consume mental resources that we need to master difficult situations. We must step back a bit to see ourselves and the people around us from a sounder distance before we make decisions. The second is crucial to understanding what motivates stakeholders and how we need to adjust our projects in order to reduce resistance and win support. If we succeed in bringing value to these stakeholders swiftly and visibly enough to be convincing to them, we may be able to change their minds and turn adversaries into supporters. Sometimes, bringing benefits to people can cost us nothing; at other times, it can be a burdensome effort regarding cost, time, and work and conflict with other goals of the project.

In all cases, if we want to run our projects with the attitude of leaders, not suppressors, we must consider the benefits that we can bring to stakeholders. Following is another example of how we can implement such an understanding in project management. This is case story that highlights the problem with ignored and rejected stakeholders. There was a similar case in the first chapter, but there, the project managers had no understanding of the attitudes of the stakeholders, as these were on the customer side and became part of the project only at a late stage. Here, the project managers in an internally performed program were aware of the stakeholders—their colleagues—right from the start, but they rejected the idea of developing connectivity with them.

From time to time, I am asked by organizations to help turn around projects. As a trainer, I am expected to know and master a wide variety of tools and practices and to be able to select those that may be most helpful in a given situation. In this secondary role, I am currently involved in a major strategic restructuring program in a group of companies that consists of about 20 individual projects. When I joined the program as an external analyst, the program and the constituting projects were suffering heavily from the resistance of employees involved. The resistance was rarely active but mostly passive, and the weak spots were generally hard to identify, but they were nevertheless strong enough to drive the program into a massive imbalance between resources invested and results achieved so far.

The organization considered the entire program to be in crisis. Each of the projects that constituted the program was late, and not one of them seemed able to deliver what was required to make the program finally successful. The program manager and the project managers firmly believed that no employee of the group would be worse off after the restructuring regarding position inside the organization, working environment, or income, but the fears of the employees in the organization were substantial. The program was very costly, and the resistance also disrupted the operations that needed to be run concurrently using essentially the same resources.

Management attention was again the most crucial resource; it was consumed by the program's woes to an unexpected degree and was therefore not sufficiently available for normal day-to-day business and its challenges. The program manager told me that she considered the subliminal fears of the corporation's employees unsubstantiated, but she predicted that further resistance from the employees would not only impact the projects and the overarching program, but also could drive the organization into a financial loss situation in which a number of those very jobs would in fact be lost.

The project managers told me that they did not want to deal with these fears, because they were unfounded and irrational. They stated that if the employees would fully understand the benefits that they would gain from the projects—including protection of their jobs—they would all support the program. When I talked with the employees in mostly private settings such as the company restaurant or the café zones, this was basically confirmed. They told me that no one from the program talked to them, that their fears were ignored—except for a virtual all-hands gathering and a static website in the intranet that did not offer any interactivity and was filled with platitudes such as, "This site is intended to help you understand the upcoming changes to the group". E-mails to the project managers remained mostly unanswered, or replies were received with preformulated, standard text that did not address the specific concerns of the original e-mails. There was no communications platform or system established that gave employees the feeling that their fears were taken seriously; instead, their impression was rather like, "Leave us alone, we have work to do in order to run our projects". Many employees had a grain of information about the projects, and when they sat together in small groups and exchanged and combined these knowledge bits, they came to conclusions regarding the projects that did not help build trust in their future but were instead deeply unsettling.

When I came back to my project managers with this information, they told me that they were fully aware of the problem, but what should they do? They were too busy to manage the delays and fix errors and did not have time to address the clandestine hostility inside the organization. Again, the urgent was the enemy of the important, and while project people had to fix too many problems at a time, they lost sight of the way to achieve the final project goals. An essential task would have been to fully on-board the employees, giving them opportunities for feedback and to communicate their concerns to the right people; but implementing such a practice would have consumed unplanned time and technical resources.

Another observation was that my talking with employees already changed things: They felt comforted that someone was there, listening to them, taking their concerns seriously and promising to pass on the message to the program manager and her team. I was not part of the decision process and was not in a position to help them directly, but my active listening brought some initial relief, upon which further actions could be built.

The corporation then arranged a number of gatherings with the employees in circles small enough to allow for questions by attendees, and the website was replaced with a portal solution that allowed open discussions with direct contributions from decision makers from the project where appropriate. Most interesting was their "Frequently Asked Questions" section, which consisted of questions that were actually asked by employees, something rather unusual for FAQs.

The investment in the engagement of the stakeholders seemed quite large, given the problems that the program had to cope with, but it was amply paid back by smoother project execution and by emotions turned from fear and distrust to positive interest and encouragement.

The program is ongoing as I write these lines. It will not be finished on time, as too many delays have added up already, but there are no new major delays adding to the existing ones, and the stress on all teams involved has significantly decreased. Also visible was a positive effect on the group's operations, which are now much less disrupted by actions based on fear of the program and are able to contribute more to the success of the organization in financial terms and otherwise.

Following the model of the three eras above, the program and its projects in the case story was moved from the geopolitical era to the connective era. When I was introduced to the situation, program and project leadership were understood as happening only inside the performing teams, including a number of subcontractors hired as consultants. The program's dependence on the employees of the organization as stakeholders was ignored, as they were not meant to be negatively impacted and therefore had no say in the projects. It was instead assumed that once the employees would see the benefits for their jobs, their support would come automatically, and, as the program manager once said, "At the end of the day, they are paid to do their job, so this is what we expect them to do". This approach may work in other organizations and other programs or projects, but it did not work here.

The project was a clear Brownfield project, and stakeholder concerns and fears must be considered in these projects as much as technical legacies. For most of the consultants, the project was a Mark n project—they had done similar projects in the past and did not see the small and marginal early signs of opposition that would finally lead the program into crisis. The consultants could not help prevent the situation either.

In the first chapter of this book, I discussed the different definitions of the term "stakeholder" over time, as it applies to investors who put tangible assets into a project, to active participants whose assets involved may be intangible, to all people, groups, or organizations whose interests may be influenced and who in turn may influence the project. With every new definition shown there, the group of people who should be considered and treated as stakeholders was expanded beyond the previous group, and it seems to me that we are in such a situation again. When one analyzes troubles that projects have with stakeholders who refuse passive acceptance or even active engagement in the project, a further extension of the definition seems necessary, adding those who perceive themselves as potentially impacted by the project or its outcomes, whether or not this is true. There may not be actual threats, but people's fears, concerns, and worries may make them stakeholders who should be identified and taken into regard. This group also includes all those who need to trust in our capabilities on technical, organizational, and interpersonal levels or will otherwise turn to fear for things they see at stake.

Leadership is a matter of perception, and so are many expectations that people have when they see our projects. When we actively identify stakeholders, the question should therefore not only be who will be influenced by the project and will be possibly influencing it, but also whose confidence and backing may we need. This question includes considering the anxieties that people may have, taking them seriously and responding to them as much as we would respond to actual impacts. Statements are not rare by project managers, project sponsors, and other individuals who assume a leadership role in a project that the resistance of people may stem from "groundless fears", from "irrational anxieties", or from their "ignorance of facts". While this may be true, motivation and engagement are also irrational, and irrational feelings often come with energy that we should redirect toward a positive course for the project. (Regarding the comment on ignorance: It is the job of the project managers to inform these stakeholders of the facts; if they fail in this task, they should not complain about behaviors caused by things people do not know.)

The dynamics of success and failure in project management pose many challenges to us that extend beyond the technical tasks that we need to perform, and also beyond the organizational and interpersonal tasks that are common in projects. We have to look at the distinctiveness of any given situation, including the people and organizations involved. We have to understand the common conflicts between particular interests and common interests, well described by game theory, and ask how we can avoid them becoming Nash Equilibria—steady and stable states to the detriment of all parties involved.

We have to look at individual people and ask what motivates them in their given life situations, and when we see that these life situations are changing, we have to ask the question again. We have to understand that in certain moments, sets of practices including behaviors, approaches, tools, and techniques may be helpful, but that they may fail in others. We should be able to classify projects and project situations and then choose the most beneficial practices for given situations, staying aware that for other situations, even in the same project, other practices may be better.

We should remain open to the fact that each project is a new learning process, and that the fundamental uniqueness of projects comes with assumptions and other kinds of uncertainties that we should actively manage. I think that one of the most basic questions for every project manager during a project should be that of the most appropriate horizon for predictions and planning. Too much agility may make us unable to cope with necessary lead times for early booking of resources and similar tasks; too much prediction may make us unable to cope with change. As leaders in our projects, we also have to give people confidence that we know the way that this project is going to take and that it makes sense for others to follow us on this way.

One may complain that these requirements make project management too difficult to handle. But they come with a positive corollary: Project management in all its complexity and unpredictability cannot be done by computers or robots. New and smarter approaches to production and administration are currently communicated under the buzz word "Industry 4.0", and there are voices that this will be the first industrialization wave that destroys more old jobs than it creates new ones. An example: During my time practicing production engineering before 1995, it was generally clear that automatic production robots were placed behind high safety fences and that humans were not allowed to enter the fenced-off areas while production was running. Today, a growing number of robots have been released from their cages. They work hand in hand with humans, and when I see these working setups, I still feel a shiver going down my spine: What will happen if the robot hurts a worker some distance from the next security cut-off switch? It is reported that the engineers of Industry 4.0 can cope with this risk, but an uneasy feeling remains.

It is unlikely that this development will negatively impact the need for project managers. Well-educated, reliable, and capable project managers will instead be needed in even higher numbers to perform just these changes to higher smartness in organizational processes, and our situational skills will be in strong demand in order to combine understanding both of the process itself and of the people involved. Computers and robots can manage many things, but they are (still?) unable to cope with unpredictable situations, and they are still not able to assume leadership positions.

Appendix A

Answers to Introductory Questions

Chapter 1

Q1 (b) For most families, a new house is a major project. They will have to agree on the configuration from the parts and will have to ensure financing. The construction company will probably regard the entire neighborhood as a project.

Q2 (a) Uncertainty on the specifications and other details of the deliverables of a project is a common element of project management. Project managers should respond to this uncertainty with methodical approaches, not with tinkering.

Q3 (d) Projects are investments, and investors sacrifice a lot to make them. They could use the money invested otherwise—for example, buy themselves a yacht, invite their loved ones to an expensive dinner, or invest the money in another project. It may take some time until the project creates benefits that may then be devalued by inflation, and each investment has a risk of loss. The results of the project fail to bring the expected benefits. Administering the investment, project managers have a responsibility to protect the investors from such damages.

Q4 (d) *Temporary* and *unique* are the attributes that distinguish projects from operations, which are continuous and repetitive by definition. A note: Operations are of course also temporary—nothing lasts forever, at least in industry. The difference is that projects are intended to be finished when work has been done and objectives achieved. Operations are ended when their processes or deliverables are no longer up to date, when resources in use are worn out, when the environment no longer supports or accepts them, or when they are no longer a sustainable and profitable business.

Q5 (b) While active risk management has become an essential element of effective project management, one should not forget that risks may occur in areas that have not been considered before, and that these risks may jeopardize even the most well-managed project.

Q6 (a) Misunderstandings are a common cause for troubled projects, and identical terminology may be used in different industries to describe different things. The danger is that people may not be aware that they are talking of different things. Project managers should develop expertise outside their industrial home domain to understand these differences.

Chapter 2

Q1 (c) Under time pressure, most people tend to rush things, but effective project managers make sure they are doing the right things, as they have no more time left to fix errors. The question relates to single-loop learning—applying corrective action and bringing the project back to plan (answer a)—and double-loop learning, which would be re-planning the project (answer d). Answer b describes a zero-loop learning approach that is sometimes also found in projects.

Q2 (a) Project managers must manage uncertainty as professionally as they manage the things they know about the project.

Q3 (d) Project managers are expected to take into account the needs and expectations of stakeholders in both internal projects and customer projects.

Q4 (b) A Nash Equilibrium is a situation in which particular interests of "players" (individuals, groups, companies, etc.) are dominant over the common interests that the players share. The emergence of a Nash Equilibrium can often be identified early by looking at the quality of communication, cooperation, risk taking, and other tangible and intangible investments that the parties make.

Q5 (a) Many organizations use the concept of the "internal customer" and "internal vendor" to model cost and profit streams inside the organization in the form of "internal charges", similar to external billing. While these models may be helpful to manage the organization, one should bear in mind that internal customers do not generate income to the organization*.

Q6 (d) If there are risk-free projects, they must be rare. Project management includes managing uncertainties, which bring risks with them. Project managers must navigate their projects through a strait with monsters at each sides. One of these monsters is called "easy"—it stands for over-optimism and beautification. The other one's name is "panic"—it is characterized by extreme pessimism and its promoters paint pictures of the future in gloomy color shades. The course between the monsters is called "realism". The two monsters have their place in organizations: "Easy" people are the best to sell their products and services, "panic" people effectively manage quality and security. Project managers stand between these groups.

---

* As management thinker Peter F. Drucker wrote: "Inside an organization, there are only cost centers. The only profit center is the customer whose check has not bounced" (Drucker, 2013, p. 84).

Chapter 3

Q1 (a)   "Over the fence" is a common term for projects that are executed along a sequence of business units or contractors, and when one of them has finished its work, it virtually throws the project "over the fence", where the next unit or contractor is waiting to do its work. This is commonly connected with a phase approach. Over-the-fence projects are highly fragmented and have no integrating function, which commonly leads to interface problems and finger pointing.

Q2 (d)   The example describes a Mark *n* project. Complacency is a risk in any Mark *n* project. The low degree of novelty makes the project look simple and easy, just a derivation of former projects. Small differences and overlooked influence factors may then surprise the team with unexpected disruptions and derail the project.

Q3 (c)   The project is a satellite project, whose success is highly dependent on the performance of the principal project (the construction of the hospital). Depending on the contractual clauses, on which project managers rarely have influence, problems of the principal project may impact the satellite project. The project manager will have to manage risks not only related to their own project but also to the principal project.

Q4 (b)   Brownfield projects take place in developed environments, in which people with their wishes, needs, expectations, fears, etc. may be a decisive factor in the dynamics of success and failure. In such projects, project managers should take special care of these stakeholders.

Q5 (d)   The question describes a composed project that is built bottom-up, not top-down, based on the preparedness of stakeholders to contribute. As a project manager, you will have to combine your skills as a facilitator and motivator with those of a coordinator, allowing the team a maximum degree of self-organization while protecting the project from running into chaos.

Q6 (a)   Applying agile principles places some burdens on the performing organization. In a customer project, it is important to have an understanding that such a burden must be shared between the customer and the contractor. It is not in the customer's interest to create distrust and frustration on the side of the contractor—whose cooperation and support is essential for project success—or to even drive the contractor into insolvency.

Chapter 4

Q1 (b)   Agile methods are great for specific projects and project situations but have weaknesses when long-term planning is beneficial to book resources early, and to order work items beforehand, and when teams cannot be collocated and dedicated to the project.

Q2 (c)   Answer (a) describes the "contributory" achieving style, answer (b) the "personal" style, and answer (d) the "entrusting" style.

Q3 (a)   "Rolling wave" is a commonly used term for iterative–incremental approaches. Another commonly used term is "progressive elaboration".

Q4 (c)   Answer (a) would build formal networks with intensive delegation, answer (b) invests own resources, and answer (d) motivates and mentors people.

Q5 (b) Agile methods are beneficial for projects with uncertain requirements and almost constant change.

Q6 (a) Even without in-depth knowledge of the Achieving Styles Model, the competitive character of this behavior should be easy to identify.

Chapter 5

Q1 (b) Assumptions registers are normally not helpful for team building but are for the other purposes.

Q2 (d) De-scoping is limited by the contract. Crashing may speed up the project, but it comes with massively increased costs and risky team dynamics. Cost engineering focuses on project costs. Benefit engineering can often (not always) find an acceptable solution for the customer to accept a higher price and postponed deadlines in return for more benefits created.

Q3 (c) The span width in StaFFA signals the distance between driving and restraining stakeholders. If this distance is large, one should expect conflicts.

Q4 (a) Coming with opportunities and threats for the project, change requests are truly two-faced.

Q5 (b) Estimates under pressure may be too optimistic or pessimistic, depending on the pressure perceived by the estimator in the given situation. Project managers should avoid forcing people to give political estimates and help them make realistic estimates.

Q6 (b) Documents and presentations shown during a kick-off meeting should be thoroughly prepared. They will be presented to senior managers who can take any weakness in them as a reason to close down your project before it has even started.

Appendix B

Traps in Terminology

There's a sign on the wall but she wants to be sure.
And you know sometimes words have two meanings.
(Led Zeppelin, Stairway to Heaven)

Misunderstandings are a common source of project problems, and most are hard to identify. Here is a list of terms and acronyms that I often found used with different meanings. Being aware of possible misunderstandings is a good first step in resolving them.

Terms and Abbreviations	Meanings	Alternative Term
Advertising (in procurement)	1. Legally mandated procedure to publish requests for seller responses in public projects. 2. Any process of finding sellers (or else) via publicly accessible media.	
Baseline	1. Project management: An approved forecast to identify and measure variances (variations) and to measure performance. 2. Configuration management: A product or service description without deltas as a means to describe change as deltas against the baseline.	

Bid	1. Seller-written offer to a prospect for price-based contract award.	
	2. Any kind of seller-written offer to a customer.	Proposal, quotation, offer, tender
Budget	1. Funding, allowance, or authorization to spend money for business activities.	Predefined budget
	2. Approved or agreed upon forecast (for example, cost baseline in earned value technique).	Budget at completion
Change management	1. A management discipline to implement changes to an organization (for example, a corporation, agency association, and so on).	
	2. An alternative term for change control or change request management. (This use of the term should be avoided.)	
Common cause	1. Universal cause of variances (variation) in a generally predictable scale.	Random cause
	2. Collective cause for two or more risks.	A form of root cause
Corrective action	1. Project management: Eliminating deviations of actual project work from the plan, while leaving the plan unchanged.	
	2. Quality management: Adjusting a continuous process when it is exceeding control limits.	
EAC (in Earned Value Technique)	1. Estimate at Completion, calculated as Budget at Completion—Cost Variance, assuming a non-recurring cost variance in the past.	
	2. Estimate at Completion, calculated as Budget at Completion/cost performance index, assuming a cost driver that influenced the past and will influence the future to the same degree.	
	3. New Estimate at Completion.	
FF	1. Network logic diagramming: Finish-to-finish dependency.	
	2. Also in network logic diagramming: Free float.	

Milestone	1. Schedule element with duration = 0, often used to highlight achievements.	
	2. Phase gate. *(This use of the term should be avoided.)*	(Phase, stage) gate, kill point, key milestone, etc.
	3. Fixed date or latest date for a delivery, go-live, handover, start of production (SoP), sign-off, and so on. *(This use of the term should be avoided.)*	Deadline, imposed date, due date
Project leader	1. Project manager on a low level, often reporting to a full project manager.	Team lead, technical project manager, project engineer, sub-project manager, etc.
	2. Project manager with strong leadership perception.	
Project resource	1. Traditional understanding in project management: (a) staff, (b) equipment and facilities, (c) material.	
	2. More contemporary understanding: Any limited assets used for a project that constrain a project team's options. May also include funding, time, know-how, and so on.	
	3. The extreme form of 2: Also includes scarce soft assets such as staff motivation, reputation, management attention.	
Proposal	1. Seller-written offer that focuses on content as well as on price.	
	2. Any kind of seller-written offer to a customer.	Bid, quotation, offer
RBS	1. Resource breakdown structure, a hierarchical decomposition of staff members and (rarely) equipment and facilities[*].	
	2. Risk breakdown structure, a tree structure that allows summing up of risk Expected Monetary Values (EMVs)[†].	
	3. Requirements breakdown structure, a hierarchical tree structure used to classify requirements[‡].	

[*] Commonly found in project management software.
[†] (PMI, 2013, pp. 317, 560).
[‡] (Wysocki, 2014a, p. 158).

Scope definition	1. *PMBOK® Guide* 3rd and following editions: Process of creating the project scope statement. 2. *PMBOK® Guide 2000*[*]: Process of developing the Work Breakdown Structure (WBS). 3. Others: Document specifying project scope. *(This use of the term should be avoided.)*	
SF	1. Network logic diagramming: Start-to-finish dependency. 2. Also in network logic diagramming: Scheduled finish date.	
SS	1. Network logic diagramming: Start-to-start dependency. 2. Also in network logic diagramming: Scheduled start date.	
Stakeholder	1. Very traditional: Investor in a project (group or individual). 2. Still traditional: Participating group or individual. 3. Contemporary: Group or individual who can exert influence on the project and its deliverables. 4. Common in public projects: Groups or individuals whose interest in the project and its environment is accepted as vested.	
TF	1. Network logic diagramming: Total float. 2. Also in network logic diagramming: Target finish date.	
Variance	1. Identified variation of actual or planned project elements (dates, costs, etc.) in comparison with a baseline. 2. In statistics: Square of Standard deviation (σ^2).	Variation, deviation

[*] (PMI, 1996).

Appendix C

What Practitioners and Experts Say

I asked a number of practitioners and experts from different locations, industries, and organizations who support the situational concept of project management for comments. I received interesting responses:

> Project Management is not about following up the written process, it is much more diverse. It's about adapting the provided tools and getting the best of them to have a successful management in our projects. We need to find what fits to our specific necessities.
>
> *Brenda Adame, Project Manager*
> *Herbalife International, Guadalajara, Mexico*

> Every project is unique; even if the final product of the project is the same as a previous project, there are changes already because people change, technology gets updated, new threats require new security measures, etc. Project management is the common framework where each role in the project can select the applicable tools, technologies, methodologies, and experiences in order to adapt them, communicate them, and apply them to the new project so it can be managed with a common goal to achieve the best possible results at the end.
>
> *Andreas Alambritis, Project Manager*
> *Easy-Forex, Limassol, Cyprus*

> How diligently a PM responds to a situation, leaving aside the standard go-by-the-book responses, not only makes the PM situationally intelligent but also makes the customer feel reassured.
>
> *Ondiappan "Ari" Arivazhagan, Chief Executive Officer*
> *International Institute of Project Management (IIPM),*
> *Chennai, India*

When clients ask me to come in as a professional project manager and consultant to solve their problems, I often face the stretch between upper management and the project team. Upper management wants the latest and best or the company-wide used standard in project management for any project. Dealing with the team is totally different. They wish me to acknowledge what they have done so far and the way they are doing things in the team. I found that I would not be successful ignoring either of the parties. So I interpret the best practice, either company or industrial standards, and shape it to the way the team functioned before. I slowly introduce methods which seem to fit nicely, whereby being very responsive to feedback from the team. That feedback is rarely in unison. There are often some in the team who need more structure and planning than others. So that is a balancing act in itself.

Andrea Behrends, President
AB&P Project Management Training, Coaching and
Consulting, Basel, Switzerland

It is often said that context trumps character. I believe that context trumps performance as well. A project is indeed a social context. At the end of the day, it is people who complete projects, not processes, and they carry them out each time in a given situation. In fact, the best process or mix of processes in the world will not lead you to project success unless there is timely and efficient communication, transparency, drive, team collaboration, and, very importantly, clear expectations. You can of course dig for the best tools and techniques and use them as needed and for the best interest of your project; however, they're just that: tools and techniques. You lead the show. Choose what's best for this team, this project, and this organization.

Radhia Benalia, Training Director
CMCS Lebanon, Beirut, Lebanon

A project is not a sprint race, but a marathon: a victory in the first stage doesn't mean anything. To win the race, "Adapt what is useful, reject what is useless, and add what is specifically your own" (Bruce Lee).

Vladimir Derunov, IT Project Manager
ARTEZIO Ltd, Moscow, Russia

One size doesn't fit all.

Elisabet Duocastella Pla, Project Manager
Self-employed, Barcelona, Spain

Oscar Wilde must have been thinking of us project managers when he said that "Consistency is the last refuge of the unimaginative". While we project managers strive to consistently deliver successful projects, we do so by adapting our methods, approaches, and actions to the situation at hand on a daily basis. We delegate the same task to two team members slightly differently, depending on their experience and background, while remaining fair to both. We adapt our communication to the needs of the receiving stakeholders while staying true to the core message. And we tailor our project management methodology to what fits best to the project and processes at hand. Being situational, adapting our approaches, and making sure that our actions best fit the current situation is at the core of what we project managers do. Oscar Wilde would be proud.

Cornelius Fichtner, Founder and President
pm-podcast.com, OSP International LLC,
Orange County, CA, USA

In our environment where supply-only projects styles are most frequent, the ability to use multiple techniques and methodologies comes from our diversely skilled team. This team functions on different aspects and segments of the project and allows each practitioner to use the methodology most suited for the specific environment. Since we often develop products and software for a specific deliverable, it is done within the overarching environment of an engineering project. Thus iterative approaches are used where required in development phases and structured approaches are used in the high-level project management.

Jacques Fouche, Head of Projects Department
Endress & Hauser, Johannesburg, South Africa

Is Agile the new and only truth for project management? Are old-fashioned sequential phase approaches outdated? There is no default answer to these questions. One has to look at the type of project to figure out the best approach. In our consulting we do not stick to the question of whether Agile is good or bad. Besides the project type or the line of business, we also look at the planning horizon.

Stavros Georgantzis, Managing Director, Founder and
Partner, The Project Group, Munich, Germany

Our team members are, at any time, willing to put in extraordinary efforts to reach our project goals and make our customers cheerful. It takes a deeply situational approach to keep the team members motivated for such a performance, and for me as the project manager, it is important that I know each team member very well.

Jörg Glunde, Service Consultant and Project Manager
Zeppelin Baumaschinen GmbH, Munich, Germany

There are several situational aspects in project management. I became mostly fascinated in software and IT projects by the need to respond in an Agile fashion to changing requirements and environmental conditions, which can later in the same project be replaced by a requirement for long-term forecasts and plans. There are methods and tools to respond to the various requirements, and project managers need help in identifying the right methods and tools for the different situations that they must master.

Herbert Gonder, Project Management Trainer
Self-employed, Munich, Germany

PM standards don't work the same way for every project—sometimes it is "just get the job done; whatever it takes" and that "whatever" becomes standard for that project.

Nasim Hossain, Architect and Project Manager
Horizon Construction and Developments, Dhaka, Bangladesh

The biggest challenge in the projects that I manage are people. Developing a collaborative spirit requires empathy and adaptive selection of practices that are favorable for the specific project situations.

Nina Jaeth, Software Implementation Manager
BMW, Munich, Germany

No two situations are the same, no two projects are the same, and the skill of the project manager is in selecting the right project management approach, applying it and keeping it under review as the project develops and circumstances change.

Nick Lake, Project Director
Green Swifts, Lindfield, UK

For the non-commercial projects of Project Managers without Borders, the social and cultural environment surrounding a project can be critical to understand. Applying standard project management practices from your own culture could lead to success, or to failure, depending on how the environment is integrated with those practices. We have to develop a deep understanding of our stakeholders before we are able to help them.

Deanna Landers, Founder and President
Project Managers Without Borders, Denver, CO, USA

We work in a very dynamic environment, where we have to interact with senior management, managers on division level, and administrative staff. There were situations in my projects that required a high degree of agility, and others when I needed to develop long-term predictions.

Srinivas Maram, Project Manager
Health Canada, Ottawa, Canada

The key to success or failure of a project must be adaptability—being frozen in a rigid framework of prescribed "best practices" cannot possibly ensure successful results in project delivery. One of the projects in the social sector that I worked with had to completely change the goals and objectives from developing a disaster relief communication system to a skills development initiative due to the legal framework in the country—situational project management where the project manager takes decisions based on the current scenario is the need of the hour.

Alakananda Rao, IT and Project Management Consultant
ALVARI Systems, Kolkata, India

A project manager must adjust the approach to the business situation: The difference between external and internal project delivery has nothing to do with whether there is an external and internal project manager. However, the pressure felt by project managers can seem very different as the external projects, those projects being undertaken for an external client, are driven directly by a fixed goal from outside so that the company that you are developing the project for will get all the possible revenue from the project once it has been completed. The aim of most internal projects is to enhance or to expand the company's business or to improve elements that exist within the company's make-up.

Mark Reeson, Project Management Advisor, Keynote Speaker
M R Project Solutions Limited, Lowestoft, United Kingdom

You can use as many PM tools or standards as you want . . . but you will mostly not finish a project on time, on budget, and of quality if you do not have the empathy to be highly responsive to changes during the life cycle of a project. This is my way for practicing situational project management.

Boris Reichenbächer, Project Manager
Daimler AG, Stuttgart, Germany

To be very clear, every project is unique, not just because of the constraints on cost, scope, and schedule, but due to the expectations and working style of customer stakeholders. Having managed more than 30 public-facing health IT projects in the past 15 years, I find that my project management tools remain consistent and effective, while it is the stakeholders that reflect the unknown of the project. Their responsiveness, dedication, quality of feedback, and communication style will dictate the adjustments I must make to ensure a productive team and successful outcome.

Christopher Scordo, Managing Director
SSI Logic, PMTraining, Walnut, CA, USA

We cannot know with certainty the outcome or consequences of decisions made today. A project is an expedition that has few prescriptions on which to rely. Projects are expeditions that evolve successfully through their adaptation to different situations. Methodology and techniques serve project management but they do not direct it. A project regime working in partnership with a "business as usual" organization requires care. Every person is unique like every project to execute. That is the beauty of this profession.

Santiago Soria, Project Manager
Sener Ingenieria y Sistemas, Madrid, Spain

Project management is contextual. "Best practices" is a misnomer. A more appropriate phrase as used in the *PMBOK® Guide* is "Generally accepted practices".

Mukund Toro, Project Manager and Consultant
Self-employed, Bengaluru, India

I was struck by the fact that, with only one exception, all of the neutral or negative impacts of achieving styles fell in the "Direct" grouping category. In other words, if you want to reduce the risk of project failure, get all the other stakeholders involved in a partnering, contributing manner!

Roger Voight, Project Manager, Project Management Trainer
Retired, Munich, Germany

The concept of there being "one best way" to achieve something was the product of scientific management in the late 18th and early 19th centuries. This idea stifles innovation and progress. Good practice and best practice are very different concepts. One of the underpinning tenets of good practice is adaptation to current needs and continuous improvement. This concept is spelled out in the *PMBOK® Guide:* "Good practice does not mean that the knowledge described should always be applied uniformly to all projects; the organization and/or the project management team is responsible for determining what is appropriate for any given project". A good methodology incorporates agility and continuous improvement by including processes for scaling and adjusting the methodology so it is adapted to be fit for purpose on each project. One size never "fits all"!

Pat Weaver, Managing Director
Mosaic Project Services Pty Ltd, Melbourne, Australia

As the largest special interest portal for project management in the German-speaking region, projektmagazin.de has a strong focus on practitioners. When we look at the various articles that we receive and publish, and also at the discussions among our 80,000 visitors per month and 21,000 subscribers, we find that project management is a fundamentally situational discipline. Project managers must be highly adaptive in selecting approaches, behaviors, tools, and methods to their ever changing environments.

Regina Wolf-Berleb, Managing Director
Projekt Magazin – Berleb Media GmbH,
München-Taufkirchen, Germany

Project management, like any other tool set, is only as effective as its selection and application. Blind application of process where it doesn't belong will often result in more harm than benefit.

Jonathan Woodcock, Project Manager
University of Waterloo, Waterloo, ON, Canada

Appendix D

Twelve Suggestions for Situational Project Managers

1. Project management is an open-skill discipline, not a closed-skill one: As a project manager, you have to respond situationally to the complex dynamics of success and failure.

2. There are no "best practices" in project management: The same practice that was successful in one project or project situation may fail in another one.

3. A fundamental feature of a project manager should be realism, which means navigating between the two monsters of "easy" and "panic".

4. The principal project resource is management attention. Having it does not guarantee the availability of resources, but without it, other resources will also be scarce.

5. Ignore stakeholders, and they will come back and bring friends, lawyers, and press with them. Turn them into Bands of Brothers and Sisters, and they will sort out your problems.

6. Actively identify, analyze, and manage assumptions to make sure that preconceptions never trump logic.

7. Actively manage risks by developing reserves. A project manager without reserves is a feeble observer of critical events.

8. The question, "What is the matrix today in my project?" is among the most important, as the answer may not be the same as yesterday.

9. Remain aware of planning horizons by asking yourself now and again how much prediction your project allows for, and how much it needs.

10. Avoid complacency and take care of the seemingly marginal details that make the project different to others and can possibly derail it.

11. Develop situational intelligence and become comfortable with a wide range of achieving styles, tools, and techniques to adjust to the specific project or project situation.

12. When you adjust your practices to changing situations, avoid the perception of jumpiness, lack of authenticity, or erraticism.

Glossary

Achieving styles	The behavioral approaches applied by individuals in leadership positions to meet requirements and attain objectives and goals.
Agile approach	An approach to managing a project lifecycle (or certain situations during a project lifecycle) with a very short planning horizon of less than month, using Agile methods to ease frequent change. Also called "adaptive approach". Many requirements are identified and described during the course of the project and "the way is made by walking". Contrasts with Waterfall and Rolling wave approaches.
Agile excuse	The defense of a lack of discipline and planning with the excuse "we are doing things the Agile way". Contrasts with the Waterfall excuse.
Agile methods	Developed originally for software development, but expanded and transferred into other application areas: Several disciplined approaches characterized by very short planning horizons, rather low planning depth, and leaderless organization. A typical aspect of Agile methods is the rejection of having a project manager and the reliance on self-organizing teams. Contrast with predictive methods.
Agilism	An ideological approach to project management that considers Agile methods generally superior to other approaches, independent of the project situation. Contrasts with Predictionism.
Balanced matrix	An overlay of one or more project team structure(s) over a functional organization structure with the project manager(s) and the functional managers on eye level. Contrasts with strong and weak matrix.
Benefit engineering	Methods to measure and positively influence the benefits from a project deliverables. May include the attempt to make a budget overrun, an increased price in a customer project, re-definition of deadlines, and other changes of fundamental parameters acceptable to stakeholders by increasing the tangible or intangible benefits for these stakeholders. Contrasts with cost engineering.

Brownfield project	A project that is performed in an already developed environment. "Brownfield" may literally refer to a piece of developed land, but the term may also be used metaphorically for similar situations in other industries.
CCB	Change Control Board. A body set up by key stakeholders to decide upon changes that are beyond the decision authorization of the team, the project manager, or the project sponsor. In a customer project, there may be an internal CCB and a joint CCB with the customer.
Change request management	A set of processes that describe how change requests will be managed in the project.
Chicken race	A game theoretical dilemma situation in which players wait for others to "jump first", for instance when they should tell a program or project manager that they are late, but hope that someone else will be late too.
CLI	The Connective Leadership Institute, a consultancy founded by Prof. Jean Lipman-Blumen to support the practical application of the Connective Leadership model.
Connective Leadership	An approach developed by Prof. Jean Lipman-Blumen to describe leadership challenges and the responses by leaders to them in a tension field between diversity and interdependence (heroism and collaboration) in a specific application of a set of nine achieving styles.
Cost engineering	Methods to measure and positively influence the costs of a project. May include attempts to identify budget overruns early and avoid and mitigate them. Contrasts with benefit engineering.
Critical incident technique	An interview technique that uses moments of special relevance for the interviewed person as an entry point to dig deeper in the person's memory. Can also be used in workshops.
Collaborative achieving style	Behavior of leaders who prefer to join forces with other people to meet tasks together.
Competitive achieving style	Behavior of leaders who prefer to compete and excel over others.
Contributory achieving style	Behavior of leaders who prefer to actively support others in achieving their goals.
Conway's law	"Organizations which design systems are constrained to produce designs which are copies of the communication structures of these organizations"[*]. Or in short: Systems reflect the relationships among those who make them.
Core team	Team members in a project who are expected and planned to be active in the project over most of its time, and who will perceive project success as personal success, and project failure as personal failure.

[*] (Conway, 1968).

Customer project	A project executed by a performing organization for a (mostly) paying customer organization. Most customer projects are profit centers. Contrasts with internal project.
Entrusting achieving style	Behavior of leaders who prefer to delegate tasks as "stretch assignments" to people who have not yet proven that they can meet them but have a potential to grow with them.
ERI	Effort-reward imbalance, a situation in which an individual considers the energy and time dedicated to the employer as much higher than the rewards (tangible and intangible) that the person receives back in return. ERI is a common cause for burn-out and for staff attrition.
Event horizon	Limitations of predictability, plannability, risk management, detectability, etc. These limitations can be fundamental or be due to inadequacies and constrained availability of resources, time, and money. Influences the planning horizon.
Factoid	A statement that is considered true by most stakeholders based on the perceived consensus. A factoid may be false because consensus is not a full replacement for evidence. Factoids are often communicated in statements such as "We all know that . . .".
Field change	Ad hoc change decision that becomes necessary during an implementation phase, often with an urgency that makes it necessary to circumvent a change request process.
Frog-king managers	Decision makers who are guided by the first sentence in the Grimm's fairytale *The Frog King*, "In olden times, when wishing did help one . . .".
Gate	A review process in a strictly sequential phase gate process, which is performed after a finished phase to review its correctness and in a second step if the project can enter the next phase. This review process can take significant time and should therefore not be confused with a milestone, which has a duration of zero.
Greenfield project	A project performed in an undeveloped environment. May be literally a green field, but may also be used metaphorically in other industries.
Hard assets	Tangible assets—money, personnel, equipment, facilities, etc.—that can be utilized by a project team as resources. These assets and the effectiveness and efficiency of their use are commonly easy to measure. Contrast with soft assets.
Hope creep	A situation when a project is late or otherwise non-performant, because team members and contractors report that they are on schedule, hoping that they can make up the lost time with overtime work, or that someone else may delay the project (chicken run).

Internal project	A project run by a performing organization for its own purposes. Most internal projects are cost centers. Contrasts with customer project.
Intrinsic achieving style	Behavior of leaders who prefer to rely on their own skills and proficiency.
IPMA	The International Project Management Association, an umbrella organization with 62 national member associations mostly in Europe, representing 40,000 members[*]. Competes with PMI for certification.
Key stakeholders	The subset of the project stakeholders who have direct and legitimate influence on the project.
Leadership	The authorization by followers given to an individual to lead them.
Mark 1 project	A first of its kind project for the project team with a high degree of novelty.
Mark n project	A project with similar predecessor projects, that give the team members confidence in their capabilities and routine.
Melon project	A project run inside an organization, where traffic-lights are used to indicate the status of a project: The melon project stands on "green" based on superficial perception, but, similar to a watermelon, the deeper one drills, the redder it gets.
Nash Equilibrium	A dilemma situation in game theory, when parties act to serve their particular interests that conflict with a common interest, and where a party must fear disadvantages when it would act to serve the common interest instead.
One-shot project	A project that gets only one chance to successfully deliver its products, services, or other kind of results.
Personal achieving style	Behavior of leaders, who prefer to build an aura of charisma around themselves.
Phantom resource	A human or non-human resource that is planned to be used in a project, but does not exist, exists but is not available, may be available but has not been formally booked, or is over allocated to more work items concurrently than the resource can handle.
Phase	A project phase is a discrete time period in a strictly sequential ("phase-gate") or overlapping ("fast-tracked") phase model. Different phases in a project may have specific teams, work contents, locations, cost centers, or other characterizing differences.

[†] (GPM, n.d., 2015).

Planning horizon	The point in the future up to which a project manager intends to plan the project. The time to the planning horizon may span anything between some days or the entire remaining duration of the project. A planning horizon may also relate to the level of detail and other aspects of a plan and is among others influenced by the event horizon.
PMBOK Guide	The *Guide to the Project Management Body of Knowledge*, a globally accepted descriptive standard for project management.
PMI	The Project Management Institute, a global professional association with over 465,000 members*. Mostly known for standardization and certification.
PMO	See Project Management Office.
PMP	Project Management Professional, a professional certification by PMI, actively held by over 665,000 individuals*.
Portfolio	In multi-project management, a portfolio is a collection of programs, projects, and operational work under a common management domain that share and compete for common resources. They may have different contents and may not share common goals.
Power achieving style	Behavior of leaders, who prefer to deploy own resources for tasks and bring order out of chaos.
Practice	Application of specific approaches, tools, techniques, behaviors, and procedures with the intention to guide action and bring about desired results.
Predictive methods	Developed originally for engineering development, but expanded and transferred into other application areas: Several disciplined approaches are characterized by long planning horizons, rather high planning depth, and organization with strong leaders. Contrast with Agile methods.
Predictionism	An ideological approach (not only) to project management that considers predictive methods generally superior to other approaches, independent of the project situation. Contrasts with Agilism.
Process assets	A combination of process know-how that an organization owns and the availability of specific resources to implement this know-how. Process assets commonly include documented procedures, forms, templates, tools, databases, and documented lessons learned. Process assets in an organization (or a portfolio of projects under one management domain) are often administered by a Project Management Office (PMO).

* By August 2015. (*Source:* Internal communications.)

Program	In multi-project management, a program is a collection of projects (and other work) that is performed to achieve common goals and benefits beyond their specific goals and benefits. The projects may be run under different management domains and may take their resources from different sources.
Project	An investment that consists of a temporary and unique set of actions performed to develop required or necessary products, services, or other kinds of results.
Project governance	A management function above the program or project manager level, mostly applied to ensure common terminology among the projects, compliance with corporate rules and processes, and legal requirements. Another goal is alignment with corporate strategy and using lessons from the projects to improve this strategy.
Project Management Office	An operational unit that manages standardized processes, procedures, templates, software, and other process assets in an organization put in place to unify methodology and terminology. This office often organizes training for the implementation of their methodologies across the organization.
Project management team	The team that supports the project manager in the tasks necessary to manage the project. It may share responsibility for project success and failure with the project manager.
Project manager	The administrator of an investment, which meets the definition of a project and has sufficient complexity to require active management under formal or informal mandate.
Ramp-up phase	A period after deliverable handover from a project to an operational production or service environment, during which the operations are performed at low rate and slowly increased, to avoid being flooded by a big number of bad results and too have resources free if initial problems need to be managed. During this time, the project shares responsibility for the deliverables with operations.
Rework	Laying hands again on a deliverable that was considered finished before to repair or alter it. Hours of rework in a project are a great metric to assess efficiency and quality in a project.
Risk finding workshop	A session of a focus group to identify risks and populate the risk register with these risks and data on them.
Rolling wave approach	An approach to managing projects with a limited prediction and planning horizon and progressive elaboration of plans. The plans are based on early descriptions of requirements that are expected to be refined and changed during the course of the project, which will then lead to refinement and change of the plans. Contrasts with Agile and Waterfall approaches.

Scrum	The most popular Agile method.
Situational intelligence	The combination of (a) the understanding that the same practice that was successful in a given situation in the past may fail in a different situation, or vice versa, (b) the ability to adjust practices to the specific needs of the project and the current situation, and (c) the care that this adaptiveness is not perceived by others as signals of lack of authenticity or reliability.
Situational project management	An approach that is based on the understanding that the same practice that was successful in one situation may fail in another one applies situational intelligence to project situations.
SitPM	See Situational project management.
Social achieving style	Behavior of leaders who prefer to delegate tasks to people who have already proven their ability to meet them.
Soft assets	Intangible assets that can be utilized by a project team as resources, such as defined processes, motivation, and reputation. These assets and the effectiveness and efficiency of their use are commonly difficult to measure. Contrast with hard assets.
SRD	Staged Response Diagram, a diagram that triggers increased attention and actions when certain limits are exceeded to protect a deadline. Can also be used to protect a budget or another kind of constraint.
Staged deliveries	The project does not have a single deliverable handover that completes the project and commences the use of the deliverables. Instead deliverables are handed over in stages, and while the team expands the scope of the product or service in steps, the team can implement feedback from the recipients (e.g., users) and incorporate it in its further development.
Stakeholder	Any person, group of people, or organization that project managers should consider during their decision processes.
Stakeholder orientation	The adjustment of decision processes to include needs, wants, expectations, requirements of stakeholders, and improve acceptance and possibly engagement by them.
Statistoid methods	Probabilistic methods that use the mathematics of statistics on activities and other items that do not have big numbers (e.g., of outputs) to validate or verify the correctness of the numbers.
Strong matrix	An overlay of one or more project team structure(s) over a functional organization structure with the project manager(s) in the more powerful position. Very common in customer projects. Contrasts with balanced and weak matrix.
Supply network	A complex system of contracts with customers and contractors that spans over three or more tiers.

Tragedy of the Commons	A game-theoretical model, in which a depletable common resource is overused by its owners, as each of the owners wishes to have a maximum personal benefit, ignoring the damage of the resource that results from the behavior. It is a form of a Nash equilibrium.
Turturism	The seemingly illogical observation that some people appear as huge and frightening from a distance but shrink to normally sized humans in proximity.
Vicarious achieving style	Behavior of leaders, who prefer to support others in achieving their goals by inspiring and motivating them.
VOC	Voice of the Customer, a concept from Japanese quality management: Customers tell the producer what they need, but in their language and with many communication deficiencies, that make it necessary to translate the information into a technically precise form.
Waterfall approach	An approach to planning and performing a project in a predictive manner with a long-term planning horizon—ideally over the entire project lifecycle. Assumes static definitions of requirements and long-term predictability and plannability. Contrasts with Agile and Rolling wave approaches.
Waterfall excuse	The rejection of an important change because it is "not in the plan". Contrasts with the Agile excuse.
Weak matrix	An overlay of one or more project team structure(s) over a functional organization structure with the managers of the functional structure in the more powerful position. Very common in internal projects. Contrasts with strong and balanced matrix.
WBS	See Work Breakdown Structure.
Work Breakdown Structure	In a decomposed project: A hierarchical decomposition of the entire project, commonly in graphical representation or in form of indented lists.
	In a composed project: A structuring system that captures the contributions of teams and consolidates them up to project level.
	In traditional application, the WBS consists of planning packages, among them control accounts and work packages, which are its lowest-level elements. In software, all WBS components are called tasks, and the lowest level may be individual activities.
Work stream	A flow of activities that are partially separate from other activities in the project and contribute to key deliverables.
Wow project	A project that is not only a good one following metrics applicable, but a great one, causing delight and excitement for stakeholders.

Zombie project	A project that is bound for failure right from the beginning, because no consideration was given to the match of project type and approach to the project, imbalance of obligations on the project with authorization and resources provided, or an environment in which other things were more important than project results.

References

Acton, J. M. and Hibbs, M. "Why Fukushima Was Preventable." (2012). Accessed on September 13, 2014, from http://carnegieendowment.org/files/fukushima.pdf.

AICPA. "Summary of the Provisions of the Sarbanes–Oxley Act of 2002." (2008). Accessed on May 14, 2015, from http://www.csus.edu/indiv/p/pforsichh/documents/03AICPASummaryoftheProvisionso ftheSarbanes-Oxley Actof2002.pdf.

Archibald, R. D. "A Global System for Categorizing Projects." (2013). Accessed on September 6, 2014, from http://russarchibald.com/global%20system%20for%20categorizing%20projects.pdf.

Argyris, C. "Inhibiting Double-Loop Learning in Business Organizations." *Reasons and Rationalizations: The Limits to Organizational Knowledge.* Oxford, UK: Oxford University Press, 2006.

Argyris, C. and Schön, D. A. *Organizational Learning: A Theory of Action Perspective.* Addison-Wesley Series on Organization Development. Boston, MA, USA: Addison-Wesley, 1978.

Arvis, J.-F., Saslavsky, D., Ojala, L., Shepherd, B., Busch, C., and Raj, A. "Connecting to Compete—Trade Logistics in the Global Economy." (2014). Accessed on March 9, 2015, from http://www. world bank.org/content/dam/Worldbank/document/Trade/LPI2014.pdf.

ASCE—American Society of Civil Engineers. Seven Wonders. (1996). Accessed on March 6, 2015, from http://cms.asce.org/Content.aspx?id=2147487305.

Axelos. *Benefits of Prince2®—The World's Most Practiced Project Management Methodology.* (2014). Accessed on January 27, 2015, from http://shop.axelos.com/gempdf/Agile_Delivering_IT_Services _contents_ and_intro.pdf.

Axelrod, R. *The Evolution of Cooperation,* Revised Edition. New York, NY, USA: Basic Books, 2006.

Barnes, K. *Volcanology: Europe's ticking time bomb.* (May, 2011). Accessed on October 21, 2015, from http://www.nature.com/news/2011/110511/full/473140a.html.

BBC News. "What Went Wrong at Heathrow's T5?" (2008). Accessed on September 10, 2015, from http:// news.bbc.co.uk/2/hi/uk_news/7322453.stm.

Beck, K. et al. "Manifesto for Agile Software Development." Accessed on August 21, 2015, from http:// agilemanifesto.org/.

Beckhusen, R. "New Documents Reveal How a 1980s Nuclear War Scare Became a Full-Blown Crisis." (2013). Accessed on August 6, 2015, from http://www.wired.com/2013/05/able-archer-scare/.

Boyle, M. and Mayes, J. L. "Set In Stone," *PM Network*. (2012). Accessed on February 28, 2015, from http://www.pmi.org/learning/settle-project-requirement-benefits-4302.

Brafman, O. and Beckstrom, R. A. *The Starfish and the Spider: The Unstoppable Power of Leaderless Organizations*. New York, NY, USA: Penguin Books, 2006.

Bretschneider, S., Marc-Aurele Jr., F., and Wu, J. "Best Practices Research: A Methodological Guide for the Perplexed." *Journal of Public Administration Research and Theory* 15, no. 3 (2005): 307–323.

BSI. *BS 6079-2:2000—British Standard—Project Management*—Part 2: Vocabulary. London, UK: British Standards Institution, 2000.

Calleam Consulting Ltd. "Case Study—Denver International Airport Baggage Handling System—An Illustration of Ineffectual Decision Making." (2008). Accessed on February 28, 2015, from http://calleam.com/WTPF/wp-content/uploads/articles/DIABaggage.pdf.

Casani, J. et al. "Report on the Loss of the Mars Polar Lander and Deep Space 2 Missions." NASA. (2000). Accessed on September 6, 2015, from ftp://ftp.hq.nasa.gov/pub/pao/reports/2000/2000_mpl_report_1.pdf.

Collin, J. and Lorentzin, D. "Plan for Supply Chain Agility at Nokia: Lessons from the Mobile Infrastructure Industry," *International Journal of Physical Distribution and Logistics Management* 36, no. 6 (2006): 418–430. Accessed on August 29, 2015, from http://www.dea.univr.it/documenti/OccorrenzaIns/matdid/matdid692585.pdf.

Collins, B. "10 Record-Breaking Bridges." (December, 2011). Accessed on March 6, 2015, from http://edition.cnn.com/2011/12/02/travel/record-breaking-bridges-bt/.

Connective Leadership Institute. "The Connective Leadership™/Achieving Styles™ Assessment Suite." (2015). Accessed on October 11, 2015, from https://www.connectiveleadership.com.

Conway, M. E. "How Do Committees Invent?" *Datamation*. (1968): 28–31. Accessed on August 11, 2015, from http://www.melconway.com/Home/pdf/committees.pdf.

Cooper, J. F. *The Pioneers*. New York, NY, USA: Wiley, 1823. Accessed on August 18, 2015, from https://www.gutenberg.org/files/2275/2275-h/2275-h.htm.

Covey, S. R. *The 7 Habits of Highly Effective People: Powerful Lessons in Personal Change*, Anniversary Edition. New York, NY, USA: Simon & Schuster, 2004.

Davies, C. "Calls for Lithium Battery Review after Boeing Dreamliner Fire at Heathrow." *The Guardian Online* (August 19, 2015). Accessed on August 20, 2015, from http://www.theguardian.com/business/2015/aug/19/lithium-battery-review-boeing-dreamliner-fire-heathrow.

Dawson, C. and Hayashi, Y. "Fateful Move Exposed Japan Plant." *The Wall Street Journal*. (July 12, 2011). Accessed on October 5, 2015, from http://www.wsj.com/news/articles/SB10001424052702303982504576425312941820794.

Deming, W. E. *Out of the Crisis*, 1st ed. Cambridge, MA, USA: MIT Press, 1986.

Drucker, P. F. *Managing in the Next Society*, 2007 ed. New York, NY, USA: Taylor & Francis, Routledge, 2013.

Fazar, W. "Program Evaluation and Review Technique." *The American Statistician* 13, no. 1 (April 1959): 9–12. Accessed on October 8, 2015, from http://www.jstor.org/stable/2682310.

Fondahl, J. W. "The History of Modern Project Management. Precedence Diagramming Methods: Origins and Early Development." *Project Management Journal* 18, no. 2 (June, 1987): 33–36. Accessed on October 8, 2015, from http://www.pmi.org/learning/precedence-diagramming-methods-origins-development-5222.

Freedman, L. *Encyclopedia of Stock Car Racing*. Santa Barbara, CA, USA: Greenwood, 2013.

Gates, D. "Boeing Shares Work, but Guards Its Secrets." *The Seattle Times* (May 15, 2007). Accessed on April 29, 2015, from http://www.seattletimes.com/business/boeing-shares-work-but-guards-its-secrets/.

Gelblum, A. et al. "Ant Groups Optimally Amplify the Effect of Transiently Informed Individuals," *Nature Communications* (July 6, 2015). Accessed on August 21, 2015, from http://dx.doi.org/10.1038/ncomms8729.

Goldoni, C. *Il Servitore di Due Padroni*. Venice, Italy: Project Gutenberg, Cambridge University Press. (1753). Accessed on March 12, 2015, from http://self.gutenberg.org/articles/servant_of_two_masters.

Goldratt, E. M. *The Critical Chain*. Great Barrington, MA, USA: North River Press, 1997.

Hachtmann, R. „Fordismus und Sklavenarbeit. Thesen zur betrieblichen Rationalisierungsbewegung 1941 bis 1944," *Potsdamer Bulletin für Zeithistorische Studien* 43–44 (2008): 21–34. Accessed on May 12, 2015, from http://www.zeithistorische-forschungen.de/sites/default/files/medien/material/2009-2/Hachtmann_aus_ZZF_Bulletin_43_44_Final.pdf.

Hammer, J. "How Berlin's Futuristic Airport Became a $6 Billion Embarrassment." (July 23, 2015). Accessed on October 19, 2015, from http://www.bloomberg.com/news/features/2015-07-23/how-berlin-s-futuristic-airport-became-a-6-billion-embarrassment.

Hardin, G. "The Tragedy of the Commons." *Science* 162, no. 3859 (December, 1968): 1243–1248. Accessed on August 18, 2015, from http://www.jstor.org/stable/1724745?origin=JSTOR-pdf.

Hart, O. and Moore, J. H. "Foundations of Incomplete Contracts." (September, 1998). Accessed on May 17, 2015 from http://eprints.lse.ac.uk/19354/1/Foundations_of_Incomplete_Contracts.pdf.

Harvey, F. and Neslen, A. "Fishing Quotas Defy Scientists' Advice." *The Guardian Online* (December 16, 2014). Accessed on August 17, 2015, from http://www.theguardian.com/environment/2014/dec/16/fishing-quotas-defy-scientists-advice.

Highlen, P. S. and Bennett, B. B. "Elite Divers and Wrestlers: A Comparison Between Open- and Closed-Skill Athletes." *Journal of Sport Psychology* 5, no. 4 (1983): 390–409.

Humble, J. and Farley, D. *Continuous Delivery: Reliable Software Releases through Build, Test, and Deployment Automation*. Boston, MA, USA: Addison-Wesley Professional, 2010.

Interbrand. "Best Global Brands." (2013). Accessed on August 29, 2015, from http://www.bestglobalbrands.com/previous-years/2013.

Johnson, R. "More Details on Today's Outage." (September 24, 2010). Accessed on October 18, 2015, from https://www.facebook.com/notes/facebook-engineering/more-details-on-todays-outage/431441338919.

Kaltenecker, S., Beck, P., Spielhofer, T., and Jaeger, S. "Agile Organization Development Manifesto." (April 10, 2010). Accessed on September 12, 2015, from http://p-a-m.org/2010/04/agile-organisation-development-manifesto/.

Kelley, J. E. and Walker, M. R. "The Origins of CPM: A Personal History," Project Management Institute. (February 1989). Accessed on August 10, 2015, from http://www.pmi.org/learning/origins-cpm-personal-history-3762?id=3762.

Kerzner, H. R. *Project Management Best Practices: Achieving Global Excellence*, 2nd ed. Hoboken, NJ, USA: Wiley, 2010.

Lehmann, O. F. "A Situational Approach to Project Management Methodologies and Behaviours Based on a Project Typology. The Case in Germany," Liverpool [Online]. (2015). Accessed on February 7, 2015, from: http://www.oliverlehmann.com/Master-Diss/SitPM-Dissertation-Lehmann-120315.pdf.

Lipman-Blumen, J. *The Connective Edge: Leading in an Interdependent World*. Hoboken, NJ, USA: Jossey-Bass Publishing, 1996.

Lipman-Blumen, J., Handley-Isaksen, A., and Leavitt, H. J. (1983) "Achieving Styles in Men and Women: A Model, an Instrument and Some Findings," in Spence, J. T. (ed.), *Achievement and Achievement Motives: Psychological and Sociological Approaches,* San Francisco, CA, USA: W. H. Freeman [Online]. Accessed February 2, 2016, from https://assess.connectiveleadership.com/articles/achieving_styles_in_men_and_women_a_model_instrument_findings.pdf.

Louven, S. *Nokia übt sich in Selbstkritik*. (November 27, 2009). Accessed on August 29, 2015, from http://www.handelsblatt.com/unternehmen/industrie/produktentwicklung-nokia-uebt-sich-in-selbstkritik/3313710.html.

Machado, A. "Latino Poemas—Antonio Machado: Caminante no Hay Camino." (August 7, 2012). Accessed on September 15, 2015, from http://www.latino-poemas.net/modules/publisher2/article.php?storyid=1115.

Mailer, N. *Marilyn*. New York, NY, USA: Grosset & Dunlap, 1973.

McGregor, J. "At Zappos, 210 Employees Decide to Leave Rather than Work with 'No Bosses.'" *The Washington Post* (May 8, 2015). Accessed on September 18, 2015, from http://www.washingtonpost.com/news/on-leadership/wp/2015/05/08/at-zappos-210-employees-decide-to-leave-rather-than-work-with-no-bosses/.

Millward Brown Optimor. "BrandZ Top 100 Most Powerful Brands Ranking." (2008). Accessed on August 29, 2015, from http://www.millwardbrown.com/docs/default-source/global-brandz-downloads/global/2008_BrandZ_Top100_Report.pdf.

Millward Brown Optimor. "BrandZ Top 100 Most Valuable Brands." (2012). Accessed on August 29, 2015, from http://www.millwardbrown.com/docs/default-source/global-brandz-downloads/global/2012_BrandZ_Top100_Chart.pdf.

Mochal, T. "10 Best Practices for Successful Project Management." (2009). Accessed on January 27, 2014, from http://www.techrepublic.com/blog/10-things/10-best-practices-for-successful-project-management/.

National Archives. "Teaching with Documents: Photographs of Lewis Hine: Documentation of Child Labor." Accessed on March 10, 2015, from http://www.archives.gov/education/lessons/hine-photos.

Nokia, "Nokia in 2002." (2003) Accessed on August 29, 2015, from http://web.lib.hse.fi/FI/yrityspalvelin/pdf/2002/Enokia2002.pdf.

Nokia. "Nokia in 2007." (2008). Accessed on August 29, 2015, from http://company.nokia.com/sites/default/files/download/05-nokia-in-2007-pdf.pdf.

Nokia. "Nokia in 2012." (2013). Accessed on August 29, 2015, from http://company.nokia.com/sites/default/files/download/nokia-in-2012-pdf.pdf.

Pennebaker, J. W. *The Secret Life of Pronouns*. New York, NY, USA: Bloomsbury Press, 2011.

Peters, T. "The Wow Project." *Fast Company*. (1999). Accessed on October 25, 2015, from http://www.fastcompany.com/36831/wow-project.

Picker, R. C. "The Razors-and-Blades Myth(s)." (2010). Accessed on March 8, 2015, from http://www.law.uchicago.edu/files/file/532-rcp-razors.pdf.

PMI. *A Guide to the Project Management Body of Knowledge—PMBOK Guide*. Newtown Square, PA, USA: PMI, The Project Management Institute, 1996.

PMI. "Case Analysis—Train Delay," Project Management Institute, p. 1 (May, 2005). Accessed on March 8, 2015, from http://www.pmi.org/learning/train-delay-impending-failure-construction-project-3237.

PMI. *A Guide to the Project Management Body of Knowledge—PMBOK Guide*, 5th ed., Newtown Square, PA, USA: PMI, The Project Management Institute, Inc, 2013.

PMI. "Requirements Management: A Core Competency for Project and Program Success." (August 2014). Accessed on September 13, 2015, from http://www.pmi.org/-/media/PDF/Knowledge%20 Center/PMI-Pulse-Requirements-Management-In-Depth-Report.ashx.

Pritsker, A. A. B. "Memorandum RM-4973-NASA: Graphical Evaluation and Review Technique." (1966). Accessed on October 9, 2015, from http://www.rand.org/content/dam/rand/pubs/research_memoranda/ 2006/RM4973.pdf.

Rosenzweig, P. M. *The Halo Effect . . . and the Eight Other Business Delusions That Deceive Managers*. New York, NY, USA: Free Press, 2007.

Sass, I. and Burbaum, U. "Damage to the Historic Town of Staufen (Germany) Caused by Geothermal Drillings Through Anhydrite-Bearing Formations." *Acta Carsologica* 2, no. 39 (February 2010): 233–245. Accessed on October 25, 2015, from http://ojs.zrc-sazu.si/carsologica/article/download/96/86.

Satariano, A. "The Phone's Secret Flights from China to Your Local Apple Store." *Bloomberg* (2013). Accessed on October 13, 2015, from http://www.bloomberg.com/news/2013-09-11/the-iphone-s-secret-flights-from-china-to-your-local-apple-store.html.

Schulte-Zurhausen, M. *Organisation*, 6th ed. München, Germany: Vahlen, 2014.

Shakespeare, W. *The Life of King Henry the Fifth, Act 4, Scene 3*. (1599). Accessed on August 21, 2015, from http://shakespeare.mit.edu/henryv/henryv.4.3.html.

Shenhar, A. and Dvir, D. *Reinventing Project Management: The Diamond Approach to Successful Growth and Innovation*. Boston, MA, USA: Harvard Business School Publishing, 2007.

Smith, A. *An Inquiry into the Nature and Causes of the Wealth of Nations*. Project Gutenberg [Ebook] (1776). Accessed on August 12, 2015, from http://www.gutenberg.org/files/3300/3300-h/3300-h. htm#link2HCH0007.

Snyder, J. R. "Modern Project Management: How Did We Get Here—Where Do We Go?" *Project Management Journal* 18, no. 1 (March 1987): 28–29. Accessed on October 8, 2015, from http://www. pmi.org/learning/origins-cpm-personal-history-3762?id=3762.

Standish Group. "The Chaos Report." (1995). Accessed on September 12, 2015, from http://www.csus. edu/indiv/v/velianitis/161/ChaosReport.pdf.

Statista. "Global Market Share Held by Nokia Smartphones from 1st Quarter 2007 to 2nd Quarter 2013." (July, 2013). Accessed on August 29, 2015, from http://www.statista.com/statistics/263438/ market-share-held-by-nokia-smartphones-since-2007/.

Statista. "Facts on Global Market Share." (2015a). Accessed on October 25, 2015, from http://www. statista.com/topics/898/global-market-share/.

Statista. "Global Market Share of the World's Largest Automobile OEMs as of August 30, 2014." (2015b). Accessed on October 25, 2015, from http://www.statista.com/statistics/316786/global-market-share-of-the-leading-automakers/.

Stephenson, A. G., et al. "Mars Climate Orbiter Mission Failure Mishap Investigation Board, Phase I Report, NASA." (1999). Accessed on September 6, 2015, from ftp://ftp.hq.nasa.gov/pub/pao/reports/ 1999/MCO_report.pdf.

Stephenson, A. G., et al. "Report on Project Management in Nasa by the Mars Climate Orbiter Mishap Investigation Board." (2000). Accessed on September 6, 2015, from http://science.ksc.nasa.gov/ mars/msp98/misc/MCO_MIB_Report.pdf.

Sterman, J. D. "System Dynamics Modelling for Project Management." Cambridge, MA, USA: Sloan School of Management, Massachusetts Institute of Technology, 1992. Accessed on October 9, 2015, from http://scripts.mit.edu/~jsterman/docs/Sterman-1992-SystemDynamicsModeling.pdf.

Sutherland, J. "Takeuchi and Nonaka: The Roots of Scrum." (October 22, 2011). Accessed on October 22, 2015, from http://www.scruminc.com/takeuchi-and-nonaka-roots-of-scrum/.

Taylor, F. W. *The Principles of Scientific Management*. New York, USA: Dover Publications, 1911. Accessed on June 19, 2014, from http://www.gutenberg.org/cache/epub/6435/pg6435.html.

The Scrum Alliance. "The 2015 State of Scrum Report." (2015). Accessed on October 9, 2015, from https://www.scrumalliance.org/landing-pages/2015-state-of-scrum-report-download.

Toyota. "Operations—The Toyota Production System." (2003). Accessed on July 1, 2014, from http://www.toyotageorgetown.com/tpsoverview.asp.

Toyota. "Jidoka—Manufacturing High-Quality Products." (2015). Accessed on October 18, 2015, from http://www.toyota-global.com/company/vision_philosophy/toyota_production_system/jidoka.html.

United Nations. "Overfishing: A Threat to Marine Biodiversity." (2004). Accessed on August 17, 2015, from http://www.un.org/events/tenstories/06/story.asp?storyID=800.

US DoD. "MIL-HDBK-881A—Department of Defense Handbook—Work Breakdown Structures for Defense Materiel Items." (July 30, 2005). Accessed on September 12, 2015, from http://129.219.40.44/adsr/VirtualMentor/documents/MIL-HDBK-881A.pdf.

Virnich, C. -J. "Der Deutsche Bauernkrieg." (March 19, 2013). Accessed on August 18, 2015, from https://www.historicum.net/themen/bauernkrieg/einfuehrung/.

Vitruvius, M. T. "Book 1: The Departments of Architecture," in *Ten Books on Architecture*, Project Gutenberg ed. Rome, Italy, ~15 BC. Accessed on March 6, 2015, from http://www.gutenberg.org/files/20239/20239-h/29239-h.htm#Page_16.

von Gerkan, M. *Black Box BER: Vom Flughafen Berlin Brandenburg und anderen Großbaustellen (= Black Box Berlin: On the Mega Airport Berlin-Brandenburg and Other Major Constructions; Own Translation)*. Cologne, Germany: Bastei Lübbe (Quadriga), 2013.

Wangler, V. *Connective Leadership, Behavioral Complexity, and Managerial Effectiveness*. Claremont, CA, USA: Claremont Graduate University, 2009.

Weaver, P. "Henry L Gantt, 1861–1919, A Retrospective View of His Work." *PM World Journal* (December, 2012). Accessed on September 16, 2015, from http://www.mosaicprojects.com.au/PDF_Papers/P158_Henry_L_Gantt.pdf.

Wood, J. C. *Henri Fayol: Critical Evaluations in Business and Management*, Volume 1. London, UK: Routledge, 2002.

Wysocki, R. K. *Effective Project Management: Traditional, Agile, Extreme*, 7th ed. [EBook]. Indianapolis, IN, USA: John Wiley & Sons, Inc, 2014a. Accessed on August 19, 2014, from http://site.ebrary.com/lib/liverpool/docDetail.action?docID=10814451.

Wysocki, R. K. *Effective Complex Project Management*. Plantation, FL, USA: J. Ross Publishing, 2014b.

Yahoo-Finance (n.d.) *Nokia Corporation (NOK)—NYSE*. Accessed on August 29, 2015, from http://finance.yahoo.com/q/hp?s=NOK&a=10&b=6&c=2002&d=10&e=6&f=2012&g=d.

Index